BIOMASS PRODUCTION AND USES

Edited by **Eduardo Jacob-Lopes**
and **Leila Queiroz Zepka**

Biomass Production and Uses
http://dx.doi.org/10.5772/58664
Edited by Eduardo Jacob-Lopes and Leila Queiroz Zepka

Contributors

Helenice De Oliveira Florentino, Dylan Jones, Daniela Renata Cantane, Margarida Vaz Pato, Aman Bouzid, Benabdeli Kheloufi, Eduardo Jacob-Lopes, Leila Queiroz Zepka, Maria Isabel Queiroz, Luis Guillermo Ramírez Mérida, Onofre Monge-Amaya, María Teresa Certucha Barragán, Francisco Javier Almendariz-Tapia, Gonzalo Mauricio Figueroa-Torres, Naoki Kano

Published by InTech

Janeza Trdine 9, 51000 Rijeka, Croatia

Publishing Process Manager Iva Lipovic
Technical Editor InTech DTP team
Cover InTech Design team

First published October, 2015
Printed in Croatia

Additional hard copies can be obtained from orders@intechopen.com

Biomass Production and Uses, Edited by Eduardo Jacob-Lopes and Leila Queiroz Zepka
p. cm.
ISBN 978-953-51-2181-7

For EU product safety concerns:
IN TECH d.o.o., Prolaz Marije Krucifikse Kozulić 3, 51000 Rijeka, Croatia,
info@intechopen.com or visit our website at intechopen.com.

Meet the editors

Prof. Eduardo Jacob-Lopes is currently Associate professor at the Department of Food Technology and Science, Federal University of Santa Maria. He received a Masters in Food Engineering in Federal University of Rio Grande, and a doctorate in Chemical Engineering from State University of Campinas. He has more than 15 years of teaching and research experience. He has published more than 300 scientific publications and 5 patents. His research interest includes environmental biotechnology with emphasis on microalgal biotechnology.

Prof. Leila Queiroz Zepka is currently Associate Professor at the Department of Food Technology and Science, Federal University of Santa Maria. She received a Masters in Food Engineering in Federal University of Rio Grande, and a doctorate in Food Science from State University of Campinas. She has more than 15 years of teaching and research experience. She has published more than 250 scientific publications and 3 patents. Her research interest includes microalgal biotechnology with emphasis on biomolecules.

Contents

Preface

Human society is facing enormous problems in the near future in order to cover the increasing demands of consumables. The current ways these demands are covered by society are not sustainable and result in unacceptable changes in our environment, such as water scarcity, depletion of natural ecosystems and global warming.

The abundance of biomass and associated favourable techno-economics has recently changed global perceptions of harnessing biomass as a valuable resource. To this end, the book "Biomass Production and Uses" aims to make a contribution to further exploring the area of biomass research and development in the form of a compilation of covering topics on production of unconventional biomasses and improving of conventional cultures, summarizing a range of useful products derived by biomass.

We are convinced that this book will be an important resource for anyone who is interested in biomasses and their technological potential, and we express the hope that it will stimulate and help researchers and industry professionals to move the field of biomass into new and improved applications.

Eduardo Jacob-Lopes
Federal University of Santa Maria,
Brasil

Leila Queiroz Zepka
Federal University of Santa Maria,
Brasil

Section 1

Biomass Production

Chapter 1

Contribution to the Assessment of Green Biomass of *Atriplex halimus* Plantation in Arid Western Algeria (Region of Naama)

Aman Bouzid and Benabdeli Kheloufi

Additional information is available at the end of the chapter

http://dx.doi.org/10.5772/61505

1. Introduction

In Algeria, the arid zones occupy a large surface area, more than 8 million hectares are used as rangelands. The ecological characteristics (soils holomorphic and superficial, very low rainfall) constitute constraints in the development of these spaces through the production of green biomass to be used as food to herds. The genus *Atriplex*, with *A. halimus*, *A. canescens* and *A. nummularia*, seems to possess the characteristics of an effective weapon against desertification, while maintaining a minimum of feed for livestock and allowing for high income to traditional feed systems [1]. The Atriplex are of great interest due to their hardiness, their good feed value, their high drought resistance and excellent productivity of wood. They can form the slicks to ensure good ground cover and can also be used as food by sheep and used as firewood and cooking in marginal areas. Also, its palatability is very satisfactory; the species *Atriplex halimus* constitutes a fodder very appreciated by all camelids, sheep and goats, particularly in times of drought [2]. It has been used in several programs of planning and restoration of degraded rangeland; or the level of excessive salinity and aridity edaphic are the main factors limiting the growth of plants and or there is also a need to provide fodder to the animal during the period of drought [3, 4]. The average annual production per hectare of the genus *Atriplex* remains relatively high compared to the other species introduced in the programs of development of dryland areas. This production is significant since it fluctuated between 11.5 and 14.9 tons of green matter per hectare, and in dry matter of 3.9 to 5.2 [5, 1, 6]. In Algeria, the use of the genus *Atriplex*, through the species *nummularia* and *halimus*, remain confined during thirty years in the development of rangeland. The main reasons which justify this situation find their sources in the needs food of herds in the arid areas where the grass is scarce. This gender is also exploited for its wood in the remote rural areas devoid of energy. All the data relating to the production of raw material palatable green and wood by the kind Atriplex justify a contribution to assess its capacity for resistance to the assaults. According to [7], rural populations are forced to exploit any available vegetation to respond to their needs in power of their herd and wood for cooking. The pastoral charge reaches more than five sheep per hectare while the opportunities are that of 0.5, which is reflected by a deterioration of plant formations or perennial herbaceous.

The average needs of rural populations for their domestic wood uses (cooking and heating) are estimated by [8] to 3 cubic meters per week. According to [9], the *Atriplex* can in certain conditions reach in biomass green supply in the order of 5 to 15 tons in Algeria and especially from 2 to 11 tons of green wood per hectare. Those figures vary considerably from one station to another, depending on the quantity of water that is actually put at the disposal of the plant and which remains a limiting factor to take into consideration because its impact is significant on the performance. In the region of Ksar Chellala (Algeria), the annual precipitation varies between 250 and 300 mm, [10] have measured the production of 2 to 5 tons of dry matter per hectare per year in the stands of *Atriplex halimus*. The plantations of *Atriplex* could therefore satisfy very quickly the real needs of local populations and would develop on all available lands. They would also probably be one of the most effective means of combating desertification and degradation as well as a considerable production of wood. The present work is a contribution to the evaluation of the production of green biomass of a plantation of *Atriplex halimus* in the region of Naama (Algeria) with the aim of controlling the temporal dimension of the production of biomass in addition to the identification of a few settings for management of Atriplexaies as the rotation duration, the period of rangeland species and productivity annual average.

2. Material and methods

2.1. Ecological characterization of the study area

The region of study is located by 33°16' N and 0° 19' W in the southern part of the west of Algeria at an altitude of 1,066 m (Figure 1). It is part of a whole geographical location called "high plains steppe" and used mainly as rangeland. The unfavourable ecological conditions (low precipitation, high thermal amplitude and sandy soils) are at the origin of the embrittlement of more and more accentuated and the reduction of pastoral ecosystems. This area belongs to the bioclimatic floor arid Mediterranean upper to cold variant with a rainfall regime of the SAWS (spring, autumn, winter, summer) seasonal type. Annually, the region of Naama receives approximately 230 mm of rain. For the follow-up period of the planting of *Atriplex halimus*, the average annual rainfall has been 219, 232 and 241 mm, respectively, for 2006–2007, 2007–2008 and 2008–2009. The average minimum temperature is of –1.1°C in the month of December and the average maximum temperature reached 35.9°C in the month of July. It is a steppe space that serves as a route to sheep breeding. The dominant soils (soils red and brown limestone) are of the steppe type, characterized by a superficial horizon of low depth (less than 25 cm), texture of sandy-loam to sandy-clay loam and slightly saline, a rate of organic matter of 0.7 % and a very low water-holding capacity, and a slightly basic pH.

For the assessment of the evolution of the production of green biomass of a plantation of *Atriplex halimus*, the choice has focused on a plantation for the value of a steppe soil bare on 500 ha (2,500 by 2,000 m) with plants of *Atriplex halimus* at a density of 2,500 plants per hectare. The seedlings are grown in the nursery of Bel Handjir in the region of Ain Sefra (same area as the perimeter of plantation), during 8 months in polyethylene bags of 9 cm in diameter and 20 cm depth. Their average height at planting was 37 cm with a main stem of 0.6 cm in diameter. The preparation has consisted of the opening of holes of planting of 40 cm depth and 30 cm side. The planting of *Atriplex halimus* was held the month of October 2006. On this plantation, three plots of 1,000 m^2 (50 by 20 m) each, far from one another 500 m according to an east–west diagonal to avoid the border effect, have been protected by a fence and monitored during 3 years. At the level of each parcel, annually have been materialized three plots of 100 m^2 (10 by 10 m) arranged in a diagonal north–south and at intervals of 10 m for a total of nine plots. An assessment of the green biomass was carried out each year in mid-October during 3 years (2007 to 2009). The average annual increase in leaf and stem has been calculated according to the following formula:

$$TAAM = \left(\sqrt[n]{\frac{final\ value}{initial\ value}} - 1 \right) \times 100$$

TAAM: (Average Annual Rate of Increase)

A **year after planting,** either in October 2007, slices at the ras of the soil have been carried out on the plots of 100 m^2 for assessing the green biomass occurred by stem used as wood for the home energy and in sheets used as livestock feed. The data have been processed by an analysis of single- factor variance organized in block, followed by a comparison of averages two to two according to the test of Newman and Keuls (software Statbox 6-4).

3. Result

The results obtained, converted to kilogram per hectare after evaluation of the green biomass of leaves and wood, are summarized in Tables 1, 2 and 3. The biomasses measured tend to increase very significantly ($p < 0.01$) in the time (Table I). The production of green biomass of leaves by planting of the *Atriplex halimus* is estimated at 5,377 kg in 3 years, resulting in an average annual increase of 1,792 kg per hectare of green leaves, or an average annual rate of 55.0%. The production of green biomass of stems is estimated at 3,358 kg, or an average annual increase of 1,119 kg per hectare, representing an annual average rate of 62.3%. The results of the biomass leaves and stems obtained are used to assess the annual average production of a plantation of *Atriplex halimus* in arid zone. This plant shows a very interesting and significant average increase compared to that of other species used for the development of rangeland.

Figure 1. Map of study site (Naama) in Algeria.

Periods of observations	October 2006	October 2007	October 2008	October 2009
Green biomass of leaves	1,125	3,913 ± 9,313	543 ± 7,348	6,502 ± 7,348
Green biomass of stems	559	1,124 ± 15,378	3,192 ± 5,385	3,917 ± 9,669
Total Biomass	1,684	5,037 ± 15,190	8,628 ± 10,5	10,419 ± 15,223

GBL: Green biomass of leaves. **GBW**: Green biomass of wood.

Table 1. Assessment of the green biomass (leaves and stems) over a period of 3 years

Type de Biomasse	October 2007	October 2008	October 2009	analysis of variance (Effect period)
GBL	3,913 [c] ± 7,511	5,436 [b] ± 4,491	6,502 [a] ± 5,909	** ($p < 0.01$)
GBW	1,124 [c] ± 9,531	3,192 [b] ± 7,439	3,917 [a] ± 5,346	** ($p < 0.01$)
GBT	5,037 [c] ± 11,529	8,628 [b] ± 8,347	10,419 [a] ± 8,981	** ($p < 0.01$)

dof: Degree of Freedom; **GBL:** Green biomass of leaves. **GBW:** Green biomass of wood.

GBT: Total Biomass. **** Probability significant at *P* <0,01**.

Table 2. Assessment of aboveground biomass produced (stems and wood in kg/ha)

Periods of observations	October 2007			October 2008			October 2009		
	GBL	GBW	GBT	GBL	GBW	GBT	GBL	GBW	GBT
1	3,901	1,120	5,021	5,448	3,199	8,647	6,512	3,927	10,439
2	3,918	1,129	5,047	5,440	3,200	8,640	6,510	3,921	10,431
3	3,921	1,112	5,033	5,429	3,193	8,622	6,501	3,901	10,402
4	3,924	1,093	5,017	5,439	3,188	8,627	6,490	3,904	10,394
5	3,912	1,148	5,060	5,438	3,187	8,625	6,497	3,911	10,408
6	3,902	1,122	5,024	5,431	3,184	8,615	6,500	3,916	10,416
7	3,906	1,127	5,033	5,426	3,195	8,621	6,495	3,925	10,420
8	3,925	1,130	5,055	5,443	3,192	8,635	6,508	3,922	10,430
9	3,908	1,135	5,043	5,430	3,190	8,620	6,505	3,926	10,431
Average	**3,913**	**1,124**	**5,037**	**5,436**	**3,192**	**8,628**	**6,502**	**3,917**	**10,419**
Average Annual Growth (AAG)	**2,788**	**565**	**3,353**	**1,523**	**2,068**	**3,591**	**1,066**	**725**	**1,791**

Table 3. Assessment of the green biomass (leaves and stems) over a period of 3 years 2007 to 2009

4. Discussion

The results obtained in green biomass both of leaves and of stems allow you to emphasize the interesting capacity for production of the species *Atriplex halimus.* The average annual increase of the green biomass total is estimated at 2,911 kg per hectare, a yearly average rate of increase of 57.7%, a value very interesting in an arid zone. According to [11] and [5], concerning production of the genus *Atriplex* ensured, according to the instalment rainfall, a dry biomass oscillating between 1,000 and 3,000 kg/ha under conditions of rainfall ranging between 100 and 400 mm. The water-use efficiency by the plant fluctuated between 10 and 20 kg of dry matter/ha per mm of rain. In Morocco, 3 years after implantation, a plantation of *Atriplex nummularia* having a density of 1,000 plants /ha has produced 1,273 kg /ha, of which nearly 31% of wood, equivalent to 625 FU and 200 kg of raw protein/ha [12]. All the authors agree to emphasize that the maximum yield of the genus *Atriplex* is around 6,500 kg of dry matter per ha under precipitation greater than 400 mm. This species is also important with regard to the production of woody biomass in extreme environmental conditions such as those of southern Morocco [13]. Compared to the other yields observed across the world (Morocco, Australia, Tunisia), the results obtained in area of Naama (Algeria) remain very appreciable and constitute a repository that can be used to justify such plantations in the high plains steppe Algerian. The experimentation carried out has allowed to determine that the duration of protection of the plantation that allows you to have the best returns is of at least 3 years for both the green biomass of leaves and stems. It follows a triennial rotation in case of use by the herds. Therefore, the duration of exploitation by cut or by release of herds in a plantation of *Atriplex halimus* in similar conditions cannot be less than 3 years.

5. Conclusion

The results obtained indicate that the capacity for growth and production of biomass and wood is very large compared to other shrubs. The average annual increase in total biomass is around 58%; In addition to its protective role of the soil and production of fodder units, *Atriplex halimus* allows to produce a woody biomass appreciable in the first year of planting. Therefore, the production of interesting green biomass and wood of *Atriplex halimus* allows to answer three basic needs in these arid spaces. The improvement, with the contribution of green biomass and a deep rooting, of physical and biological characteristics of land threatened by desertification and salinization, the availability of a green biomass serving palatable feed for the herd and the use of wood as domestic energy renewable. Knowing that all the plantations of *Atriplex* in Algeria are subject to exploitation by the herds in search of fodder units rare in the arid areas for the first year, it appears urgent to proceed to the prohibition of grazing for 3 years to ensure better yields. This prohibition of grazing would also respond to the needs in wood for cooking and heating of the nomadic populations and riparian plantings, using the stems of *Atriplex halimus* which are not consumed by the sheeps. In the light of the very encouraging results, the use of *Atriplex halimus* could be generalized under the climatic and edaphic conditions similar to those of the experiment since it allows a supplementary feeding forage and a source of renewable energy for heating and cooking in semi-desert area.

Author details

Aman Bouzid* and Benabdeli Kheloufi

*Address all correspondence to: aman_bouzid@yahoo.fr

Faculty of Natural and Life Sciences, University Abdelhamid Ibn Badis of Mostaganem, Mostaganem, Algeria

References

[1] Le Houérou HN. Utilization of fodder trees and shrubs in the arid and semi arid zones of west Asia and North Africa. Arid Soil Res Rehab 2000;14:101–35.

[2] Kinet JM, Benrebiha F, Bouzid S, Laihacar S, Dutuit P. Biodiversity study on Atriplex halimus for in vitro and in vivo detection of plants resistant to harsh environmental conditions and for potential micropropagation. Cahier Agricult 1998;7:505–9.

[3] Le Houérou HN. The role of saltbushes (Atriplex spp.) in arid grazing land rehabilitation in the Mediterranean basin: a review. Agroforestry Syst 1992;18:107–48.

[4] Bajji M, Kinet JM, Lutts S. Osmotic and ionic effects of NaCl on germination, early seeding growth, and ion content of *Atriplex halimus* (Chenopodiaceae). Can J Bot 2002;80:297–304.

[5] Correal A, Otal J, Belmonte C, Sotomayor JA. Evolution of sheep production under continuous rotation grazing of a salt bush plantation of Atriplex sp. in south-east Spain. In: 4th Int. Rangeland Congress. Montpellier, France. 1990;1:422–25.

[6] Cherouk A. Contribution to the definition of a methodology to estimate the biomass *of Atriplex nummularia* in eastern Morocco. In: Gintzburge G, Bounejmate M, Nefzaoui A. (Eds.) Fodder shrub development in arid and semi-arid zones. Proceedings of the workshop on native and exotic fodder shrubs in arid and semi-arid zones, 27 October to 2 November 1996, Hammamet, Tunisia. Vol. II. ICARDA, Aleppo (Syria) 2000, pp. 334–339.

[7] Benabdeli K. Aspects physiognomic-structure of the woody vegetation facing the anthropologenetic pressure in western Algeria. PhD thesis of State, University of Sidi Bel Abbes, 1996.

[8] Benabdeli K. Methodology for assessing the impact of man and animal on the woody formations. PhD thesis of specialty, University of Aix-Marseille II, 1983.

[9] Ouadah Y. Contribution to the study of the main species of interest feed of arid and semi-arid regions of Algeria. Application to a few species. Memory of end of study, INA, Algiers, 1982.

[10] Rodin LE, Botschantzen V, Kalenov H, Miroshnichenco YM, Pelt N, Vinogradov B. Geo-botanical studies the pastures of the western sector of the department of the Medea (Algeria). Naouka Ed., Leningrad, 1970.

[11] Forti M. Introduction of fodder shrubs and their evaluation for use in the semiarid areas of the northwest Neveg. Institute for Arid Zones Research, Bear- Sheva, Israel. Minco, 1971.

[12] El Mourid M, Malki M, Sbeita A, Chiriyaa A, Nefzaoui A, Shideed K, Haj Hass an Aawawedha F, Sweidan SY. Crop livestock integration: alternatives to stop desertification in arid regions: WANA expert meeting: scientific research and its role in combating desertification and stabilizing sand dunes. Taghit, Algeria November, 2001, 4–6.

[13] Benchaabane A. *Atriplex halimus* L.: source of firewood and fodder in arid environments (case of the region of Marrakech, Morocco). In: Gintzburge G, Bounejmate M, Nefzaoui A. (Eds.). Fodder shrub development in arid and semi-arid zones. Proceedings of the workshop on native and exotic fodder shrubs in arid and semi-arid zones, 27 October to 2 November 1996, Hammamet, Tunisia. Vol. II. ICARDA, Aleppo (Syria) 2000, pp.610–616.

Chapter 2

Production and Management of Sugarcane Biomass — Process Optimization

Helenice de Oliveira Florentino, Margarida Vaz Pato, Dylan Jones and Daniela Renata Cantane

Additional information is available at the end of the chapter

http://dx.doi.org/10.5772/61533

1. Introduction

The worldwide search for alternative fuels for energy purposes has been growing for environmental and economic reasons. One of the renewable resources of considerable interest is biomass.

In countries where there is a vast area of fertile land, a tropical climate, and available water resources, the growing of sugarcane for subsequent use of its biomass is a viable alternative [35]. From the sugarcane, one may produce various types of sugar and alcohol as well as generate electricity. Even after processing its agricultural and industrial residues such as straws, bagasse, stalks, pulp, molasses, filter cake, and other products, it can be further exploited. The predominate products derived from the sugarcane as potential alternative energy sources are ethanol alcohol, bagasse, and residue of harvesting.

On the other hand, at present, there are significant logistical problems associated with producing biomass from sugarcane. The huge size of the agricultural area and the combined management of the economic, technical, political, social, and environmental factors cause the management of the the sugar–ethanol mills to become extremely complex. Thus, decision makers have sought help by resorting to mathematical and computational tools in an effort to optimize decision making in a safe, economic, and environmentally correct manner, thus making the use of this biomass more appealing.

Section 2 of this chapter addresses the major problems of the sugarcane crop, which may be solved by resorting to optimization techniques, while Section 3 discusses mathematical modeling and solution approaches to these problems.

Figure 1. Source: https://www.flickr.com

2. Improving the quality, production, and management of the sugarcane biomass

The sugarcane is a plant belonging to the genus *Saccharum* L, of which there are at least six species. The sugarcane planted today is a multispecific hybrid called *Saccharum spp*. It is a semiperennial crop since after planting it is cut several times before being replanted. The sugarcane species comes from South East Asia and today has become one of the leading world crops, grown in over 100 countries. About 80% of the planet's production is concentrated in ten countries, with Brazil and India accounting for over half the world's output. On average, its productive cycle ranges over 6 years with five harvests.

The productive cycle of sugarcane begins with preparation of the land and the subsequent choice of the variety to be planted. The cycle then involves development, ripening, and harvesting. The sugarcane resprouts and follows the cycle once more and can be cut up to five times. Then replanting takes place and the entire procedure repeats itself. However, whenever it is replanted, the sugarcane's quality suffers significantly in terms of biomass production, sucrose output, and fiber, in other words, in every measure that directly affects the end products: sugar, alcohol, and energy. Thus, considerable investment is required in the factors that improve cost, production, and quality of the cane without affecting environmental sustainability. Key factors include appropriate management, choice of the optimal variety to be planted, soil preparation, and the correct period for planting and harvesting.

The management of all these factors is extremely complex, and the farmers and factory managers have to resort to various tools to assist them. In this context, optimization techniques have proved to be fairly efficient, as they can contribute to optimized decision making, the achievement of optimum production goals, and the planning of environmentally sustainable processes.

The literature presents a vast amount of problems in which the solution can be assisted by using deterministic optimization methodologies. The application of these methodologies consists initially in studying the problem followed by its mathematical modeling, using the knowledge acquired on the issue concerned to identify parameters, variables, objectives, and constraints. Each optimization model has specific characteristics that depend on the problem that generated it. Functions defining objective function(s) and constraints may involve nonlinearities, convexities, and other particularities. Depending on these characteristics and the required solution technique, the model will be classified within an area of optimization. The next step is to validate the model by studying its performance and coherence of its mathematical results in relation to the reality being modeled. There then follows a discussion of the solution reached with specialists from the field.

Section 2.1 presents the main factors that characterize sugarcane biomass. Sections 2.2 to 2.5 discuss four problems which may be solved with the help of deterministic optimization techniques.

2.1. Sugarcane quality

The method of assessing the quality of the raw material should align with the needs of the producer in order to grow an economically viable crop while meeting the industry requirements. This implies that the choice of new varieties developed through genetic improvement should meet all these prerequisites.

According to Engelke [9], several measures may be used to assess the quality of sugarcane. These are the percentage of pol (sucrose) in juice and in cane, percentage of brix (total soluble solids) in juice and in cane, percentage of fiber in cane, commercial cane sugar (CCS), and purity. The process of determining sugarcane quality and calculation of each of the above measures is explained in detail by Engelke [9].

Among the measures detailed to ascertain the quality of the cane, the ones respecting the fiber and sugar contents should be emphasized. Fiber is a material that is insoluble in the water contained in the cane and constitutes a substance of great importance in the sugarcane agroindustry. From an agricultural standpoint, the varieties that are richest in fiber resist falling more easily, even when they are subject to straw removal and fire and are generally more resistant to the intrusion of pests in the stalk. From an industrial stance, the amount of fiber is important for the industry's energy balance since the fibers (bagasse and pulp) can be used for burning in boilers, generating steam which is converted into electric energy for the mill itself. At the beginning of the harvest, it is of fundamental importance that the varieties have greater fiber content to ensure the provision of fuel for the boilers. Another significant measure of the sugarcane's quality is the amount of sugars (pol percentage in cane) as it is mainly from sucrose (for sugar production), glucose, and fructose (for alcohol production) that the industrial output of sugar and alcohol stems. The level of sucrose is directly related to the point of sugarcane maturation since at this point the levels of sucrose are highest. Besides the content of sucrose, the point of sugarcane maturation is determined by the content of reducing sugars (glucose and fructose) and pulp humidity during the crop's cycle [2, 7, 27].

The period of time in which the variety displays appropriate technological conditions to be harvested is the useful period of industrialization, known as the period of industrial utilization (PIU). The PIU commences when the sugarcane reaches the content of sucrose (pol percentage in cane) above the accepted standard (in Brazil it is above 12.257%). From this point, it continues to increase and once it reaches its peak, the sugar content starts decreasing to a minimum permitted figure. This limit is imposed to avoid a significant loss in sugar content. In Brazil, the end of the PIU is determined when the sugar content falls to the minimum of 16%. Different varieties present different values of PIU, some shorter and others longer. Hence, the knowledge of the range of this period is extremely important when it comes to planning a variety's planting and harvesting. In this way, one can make the most of the quality of the sugarcane [7, 35]. Figure 2 illustrates the PIU values of varieties with different maturity curves.

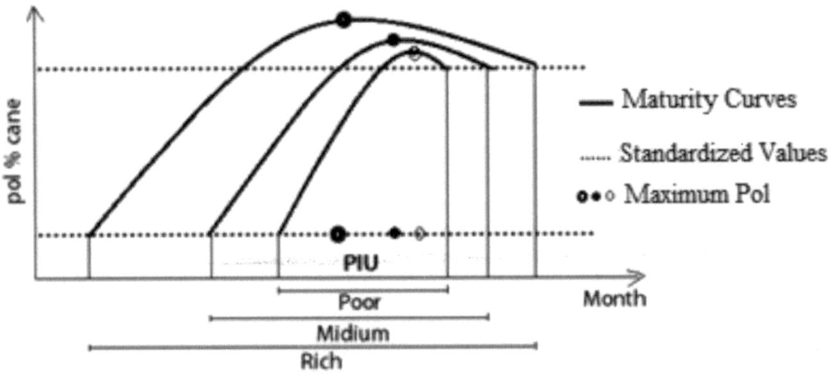

Figure 2. Maturity curves for poor (with a short PIU), medium (with a medium PIU), and rich (with a long PIU) cultivars [35].

In Brazil, the varieties regarded as poor display an appropriate period for harvesting of approximately 3 months. The varieties regarded as medium enjoy a period designed for a more flexible harvest, with a period of up to 4 months. The rich varieties can be gathered with maximum harvest flexibility, extending the period to up to 6 months, depending on the cultivar [35].

The fiber and pol contents are important means of measuring cane quality, but there are also other important factors to ensure the success of the sugar cane plantation in the production of quality biomass, such as preparation of the land for cane planting, choice of the right period for planting and harvesting, and the time the cut cane waits for milling [33].

On the basis of the quality factors discussed, mathematical optimization methodologies can help to determine the planning processes for sugarcane by striving to increase its quality and/or quantity.

2.2. Preparing the land for planting

Preparing the land for planting has a direct influence on the production of sugarcane and hence assures economic, social, and environmental benefits for the sugar–alcohol sector. Because a favorable utilization of the area implies an increase in cane production, improvements in transport logistics, simplification of planting and harvesting, control of pests and weeds, and minimization of the number of maneuvers of the harvesters lead to reductions in the use of pollutants and fuel costs [16, 22].

Figure 3. Source: https://www.flickr.com

To profit more from the available area, a planning process is needed that considers the shape of the plots and the layout of the tracks, in keeping with the area's relief and soil. Thought should be given to the length and width of the plots in the light of the slope, as every terrace should have a track. The tracks should be rationalized, where the area used for the path system should range between 2.5% and 4.0%. It is recommended that secondary paths are made to adequately accommodate the use of machinery. Early road planning is important to define the road network and consequently the shape of the plots and position of the furrowing [4, 16, 22].

The best recommended plots are those that are rectangular in shape and as long as possible. These are compatible with highly efficient paths that avoid sharp bends and the need for the harvesters to make excessive maneuvers, leading to a loss in time and fuel consumption. One should choose an area devoid of trees, fences, stumps, and large stones, which may prevent the movement of machines. A general cleaning up should be made, removing all the irregularities such as furrows, gullies, and holes. The degree of compression of the soil and its depth should be detected, principally in areas of reform. Besides this, the plots should not have slopes greater than that which is permitted for the machinery available in the market [1].

Problem 1: Optimized partitioning of the land into plots.

Optimized planning of the partitioning of land into plots consists of dividing the area available for sugarcane planting into regular plots of adequate sizes, so as to increase the ease of turnover of the cane harvester and hence to minimize the fuel cost arising from the harvesters' maneuvers.

After preparing the soil and partitioning the land into plots, one should make the right choice of the variety to be planted in each plot, so as to enhance the production of the cane in terms of quality and quantity.

2.3. Selection of the sugarcane varieties

Among the sugarcane's production and quality factors, the choice of the variety to be planted constitutes the most important decision since it provides the basis of the other technologies of production and processing of raw materials, thus providing significant increases in industrial productivity without an increase in production costs [12].

Figure 4. Source: https://www.flickr.com

The varieties should display desirable features such as high yield, high sugar content, capacity to reshoot, no tendency to fall, and resistance to pests and diseases. The moment one chooses one variety to be planted, one must note its features and adaptation to the environment; otherwise, the productive and quality potentials may be compromised [29].

The right choice of sugarcane variety for cultivation is not an easy task, as it depends on a host of fundamental information related to agronomic and industrial factors, besides the interaction of biotic, abiotic, administrative, and economic factors. Many studies have been performed with a view to proposing mathematical optimization models to assist this choice.

Problem 2: Selection of sugarcane varieties to be planted.

The problem consists of determining which are the sugarcane varieties best adapted to local soil and climate such that they can be selected to be planted in the available plots, respecting constraints and optimizing the relevant objectives. The areas of the plots as well as the distances of the plots to the mill are generally known. The objectives include cost, profit, or production optimization and energy balance. Constraints ensure sugar and fiber demand is met as well as placing bounds on the planting areas, either in total or per variety.

2.4. Sugarcane cultivation and delivery

The sugarcane cultivation cycle extends from the preparation of the soil to delivery of the cane at the factory. Among the important stages of this cycle, planting and harvesting play a major role. In fact, planned planting and harvesting lead to a series of benefits throughout the cultivation cycle and in the industry. These are improved utilization of the area, increased cane output, improved transport logistics and better cane reception in the factory, administrative simplification of the industrial activities, enhanced response to the demands of the industry, improved utilization of labor, enhanced cost planning, and improved control of pests and weeds, among others.

Figure 5. Source: https://www.flickr.com

Setting up a sugarcane plantation involves a series of factors as it is semiperennial. To ensure that the harvest, mainly mechanized, is successful, attention must be paid to the planting process as the lifespan of the cane plantation depends on the interaction of the two operations detailed in previous sections. Regarding the first operation, many factors determine the quality of planting from its density, soil preparation, planting season, choice of variety, quality, and age of the plant and parallel alignment of the rows of plants. Quality planting directly influences not only the factors that determine if the crop will ensure a high yield following harvest, but also the reduction of crop production costs. The importance of the early operations calls for good planning and considerable technical know-how. In fact, the planting process involves high costs, and decision making in this phase will influence the entire plantation cycle [23, 32].

The harvest season is determined by considering certain factors, such as the amounts of sucrose and reducing sugars contained in the cane juice. For this reason, it is not recommended to begin harvesting until the canes have achieved the minimum technological standards for industrialization nor delay the harvesting season to avoid a fall in the average yield. The output environment may ensure the sugarcane crop is handled better, thus guaranteeing its maximum economic exploitation [23, 32].

The principal indicator in determining the moment to renew a cane plantation is related to the yield noted in the course of its ageing. The greater the number of cuts, the greater the chances of a fall in productivity and the greater the likelihood of the need for renewal. Sugarcane farming is more productive in the first cuts, with a strong tendency for the yields to decline as the years they remain standing in the plots [7, 23, 32, 35].

One problem encountered in high-yield sugarcane regions is that of ensuring that the cane harvests in all plots are performed as closely as possible to the date at which the cane has accumulated the highest PIU value (maximum point on the maturity curve, see Figure 1). Should this be possible, then one may make the most of the biomass to produce sugar and alcohol and to generate energy as well as avoid the need for chemical products. On the other hand, this is an extremely difficult task, as the factories have fixed transport and sugarcane milling capacities and have to meet demand in every period of the year. Often, out of necessity, the cane is cut outside the PIU due to factory demand, industrial limitations, or to avoid the cold periods in zones where the cane freezes, leading to high biomass loss [7]. To overcome this problem, one must undertake optimized planning of sugarcane planting and harvesting.

Problem 3: Planning the planting and harvesting of sugarcane.

Optimized planning of planting consists in deciding in which periods the cane should be planted in each plot so as to satisfy demand during every period of the subsequent 4 or 5 years. The planning of the harvest is linked to the planning of the planting, as planting in each plot should be performed in such a way that the cane harvest is undertaken close to the date of the maximum pol cane content while satisfying demand and respecting the mill's technical limitations. In this sense, mathematical optimization modeling can be of considerable assistance in producing the combining planting and harvesting plans.

2.5. Utilization of sugarcane residual biomass

Mechanized sugarcane harvesting generates large amounts of residue. This residue, known as harvesting residual biomass, is made up of stalks, straw, leaves, and fragments of cane stems. The purpose of this biomass has been widely studied, and various research papers demonstrate the feasibility of using stalks to produce energy in view of the high useful calorific value of this material. The calorific value varies according to the growing conditions and the type of variety [2, 3, 28, 30]. The process of handling the harvesting residual biomass is very costly, and the advantages of cropping, recovery, and utilization have mobilized university researchers, factory managers, and directors, all of whom are keen to find the most productive, economical, and effective way to conduct this operation. Biomass-related factors may provide another consideration in the choice of varieties. Optimized biomass allows for improved economic results and helps to implant the residue in the mill's energy production system. According to Sartori et al. [29], in an attempt to minimize the environmental impact and the influences affecting productivity, and hence the sugar–ethanol mill's profitability, researchers have persisted in choosing the sugarcane variety that produces stalks with a higher calorific value and a low harvesting cost without losing its yield features. Only in this manner will it be feasible to utilize this residue for energy co-generation.

Figure 6. Source: https://www.flickr.com

Problem 4: Utilization of residual biomass to generate energy.

The problem consists in the determination of the optimized process to collect residual biomass from the field (straw) and from the industrial process (bagasse) to be used for energy generation purposes.

Such processes may comply with the objectives of minimizing the cost of harvesting and transportation of the biomass to a processing center, or maximizing the energy and economic balance in the sugarcane processes, or other economic or environmental objectives.

3. Optimization processes

In the light of this biomass profile for the generation of energy, sugarcane has become one of the most important crops, and considerable attention has been devoted to its cultivation. It is therefore fundamental to assess its renewability and sustainability, aiming to improve the raw material intended for the sector. This includes the complete production system from soil preparation through to extraction of the raw material leading to its ultimate use. Due to its dimension and complexity, the productive sugarcane chain is faced with a range of different problems. Therefore, various tools that may help to solve these problems have been used, of which one of the most important is optimization modeling.

3.1. Mathematical programming models

Optimization is a process in which one strives to obtain the best combination of several factors (decision variables), given certain limitations (set of constraints) in keeping with one or more particular objectives (objective functions).

A general problem of single-objective optimization consists in minimizing or maximizing a function, where its domain, when dealing with mathematical programming, is represented by the set $\{ x \in R^n \,/\, g(x) \leq 0\}$, which could also be expressed through the conditions $g(x) = 0$ or $g(x) \geq 0$:

$$\text{Minimize } f(x) \ (\text{or maximize}) \tag{1}$$

subject to

$$g(x) \leq 0 \tag{2}$$

$$x \in R^n \tag{3}$$

where x is the vector of decision variables; function f: $R^n \rightarrow R$ is known as the objective function; a set of constraints is defined through functions g: $R^n \rightarrow R^p$. These define the feasible region, i.e., the set of feasible solutions, $n > 0$ and $p > 0$, are integer numbers associated, respectively, with the number of decision variables and constraints.

Depending on the nature of the optimization problem (Equations (1)–(3)), the objective function, constraints, and variables assume different characteristics, calling for different solution techniques. These different problems are classified according to the characteristics of the objective function, equations, or inequalities, which describe the restrictions and the decision variables:

- If $f(x)$ and $g(x)$ are linear functions, and $x = (x_1, x_2, ...x_n)$, $x_i \in R$ for $i = 1, 2,..., n$, one has a linear programming problem (LP).
- If $f(x)$ and $g(x)$ are linear functions, and $x = (x_1, x_2, ...x_n)$, x_i integer for $i = 1, 2,..., n$, one has an integer linear programming problem (ILP).
- If $f(x)$ and $g(x)$ are linear functions, and $x = (x_1, x_2, ...x_n)$, x_i integer for some but not all $i \in \{1, 2,..., n\}$, one has a mixed integer linear programming problem (MILP).
- If $f(x)$ and/or $g(x)$ are nonlinear functions, and $x = (x_1, x_2, ...x_n)$, x_i integer for $i = 1, 2,..., n$, one has an integer nonlinear programming problem (INLP).
- If $f(x)$ and/or $g(x)$ are nonlinear functions, and $x = (x_1, x_2, ...x_n)$, $x_i \in R$ for $i = 1, 2,..., n$, one has a nonlinear programming problem (NLP).

The counterparts of these problems with binary (0 or 1) instead of integer variables have also been widely used. The corresponding acronyms are BLP, MBLP, and BNLP. If there exist q objectives $f_1(x)$, $f_2(x)$, ..., $f_q(x)$ to be maximized or minimized rather than a single objective, then one has a multiobjective programming problem (MOP). If the multiple objectives have target values to be achieved rather than being purely of the minimization or maximization type, then one has a goal programming problem (GP). Both multiobjective and goal programming models follow the same classification conventions with respect to integer variables and non-linear functions as the single objective case.

There are many other varieties of optimization problems, depending on the characteristics of the functions and variables involved [17, 18, 19]. For each class of problems, there are one or more exact or approximate specific solution methods [5, 8, 24, 36].

3.2. Approaches within sugarcane processes

The quality and quantity of biomass to be produced and the activities involving growing, harvesting, transportation, processing, and commercialization of the sugarcane are factors that may be optimized with the help of optimization techniques. Several studies dedicated to optimization models to resolve the above-mentioned Problems 1, 2, 3, and 4 have been published in recent years.

Problem 1: Optimized partitioning of the land into plots.

Consider an available area for planting sugarcane in a field. k is the number of possible plots that can be allocated to sugarcane in this area. Cherry et al. [6] defined the plot generation problem as follows. The planting area must be partitioned into rectangular plots with dimensions (l_j, w_j), where l_j is the length and w_j is the width of the plot j $(j=1,2,...,k)$ in order to increase yield, reduce traffic, and minimize the maneuvers of the sugarcane harvesting machines while respecting all the constraints imposed by mill.

According to Cherry et al. [6], the planning begins with soil preparation and the partitioning of the planting area into sugarcane plots. The main feature of plots is that they must be rectangular to prevent excessive maneuvers by the harvesting machines. The cited authors propose a methodology using an NLP model for planning the division of the plantation area into plots in order to perform mechanized harvesting. As the plots are rectangular, the authors used a two-dimensional cutting theory based method to solve the problem. Computational experiments were performed regarding real cases, and the proposed methodology shows a reduction of over 40% in the number of maneuvers of the sugarcane harvesting machines, thus implying many economic and environmental advantages.

Problem 2: Selection of sugarcane varieties to be planted.

This problem consists in deciding which of the n varieties of sugarcane, adaptable to local climate and soil, should be planted in each of the k plots, with size L_j, and distance from the cane's processing center given by D_j $(j=1,2,...,k)$, in such a way that it optimizes one or more objectives, whether it be to minimize costs and/or to maximize production, maximize profit, or others. The solution should meet the company's recommendations to maintain cane quality and the demand for sugar and alcohol. Examples of these constraints include the limitation of the average sucrose and cane fiber content and the utilization of the entire area set aside for the sugarcane plantation.

Sartori et al. [27] proposed two LP models for this problem. The first model involved the selection of varieties of sugarcane to be planted meeting the mill requirements to minimize the quantity of residue produced. The second model discussed the use of residue to produce energy. This is related to the selection of varieties and quantities to be planted in order to meet the requirements of the mill, to reduce the quantity of residue and to maximize the energy production. The models developed permit the optimization of the energy available in sugarcane residue and its quantity, with the purpose of selecting the best adapted varieties for the production of energy from the biomass or for the production of compost. With these models, it is also possible to determine the area to be planted per variety, the amount of pol to be produced, the amount of residue, and the amount of energy to be extracted.

Florentino and Sartori [13] linked the two problems proposed by Sartori et al. [29] with a BLP model to support variety selection and planting quantity of sugarcane in order to reduce crop residue, maximize energy generated by this residue, and satisfy the demand of the mill. They solved the conflict between these objectives by using nonzero-sum game theory. One player was associated with residue and another with energy. From the Nash equilibrium points supplied by the game, it was possible to choose a solution that satisfies the mill's interests, thus reducing the sugarcane crop residue and increasing the energy generated by this residue.

Sartori and Florentino [28] presented a BLP approach to a problem that is similar to the ones discussed by Sartori et al. [29]. The approach does not involve determination of the planting area per sugarcane variety but instead focuses on the decision about the variety to plant in each plot. The model of [28] is more realistic since in practice there is only one variety per plot.

Piewthongngam et al. [25] proposed an optimization model for planning cultivation of sugarcane by selecting the time and the varieties that each producer in the Northeast of Thailand should plant, avoiding the generation of excess supply during the peak of harvest. The planning takes place over a long time period and determines the cultivation period, the varieties to be planted, and the time windows of harvest for each farm so that the total sugar production is optimized. The proposed LP model allows decision makers to visualize the sugarcane production in each farm individually on different dates and with different varieties. The results presented by the authors using mathematical programming showed a potential increase of the 23% in sugar production when compared with the traditional planning method.

Florentino et al. [11] proposed a multiobjective ILP model to choose sugarcane varieties so as to minimize costs in the use of crop residue and simultaneously to maximize the energy balance. The model assists the selection of planting varieties by supplying the lowest costs for transferring residual biomass from the harvest in the field to the production center and the optimized residual energy balance. It thus provides the mean energy and fiber content of the varieties of sugarcane selected for planting, taking into account the mill's requirements. The above-mentioned authors encountered difficulty in solving this model using exact methods for large-sized instances. The two works mentioned below followed up [11] by trying to remedy these difficulties. Thus, Homem et al. [15] used a hybrid procedure involving the primal–dual interior point and the branch-and-bound method to solve the problem. The methodology presented a good computational performance and produced reliable practical solutions, but only for small size problems. Florentino and Pato [10] studied the computational complexity of the problem and showed that it is NP-hard. They proposed a solution methodology using a bi-objective genetic algorithm. A computational experiment undertaken with a set including real and semirandomly generated instances was reported, thus showing the practicality of the technique.

Problem 3: Planning the planting and harvesting of sugarcane.

The problem consists in determining the period of the year in which the cane should be planted and harvested in each plot over four consecutive years, so as to maximize the total cane production over the planning horizon. Constraints should be respected, such as imposing the guarantee that the cane be planted in all plots in the first year, a single variety be planted per plot, the guarantee of meeting the mill's pol and fiber demands every planning year and the guarantee that the factory's cane milling capacity be satisfied in all the harvest periods.

Milan et al. [21] presented an ILP model to minimize the cost of transportation of sugarcane from field to mill by integrating road and rail transport systems. The model presents constraints related to the continuous supply of sugarcane in the mill, time that the harvesters can work, type and capacities of vehicles for transportation, and the storage capacity of sugarcane and the availability of routes. According to the authors, the results showed that the model is useful for minimizing the cost of transportation and also for scheduling the transportation of sugarcane, even with the large number of variables and constraints that are present in the model.

The transportation logistics during the sugarcane harvest process is a difficult problem for mill managers to solve. Higgins [14] formulated and implemented an MILP model to assist in resolving operational problems and costs of transportation of sugarcane in Australia. The model improved the scheduling of vehicles and thereby reduced the number of vehicles needed as well as the queues and downtime of vehicles at the mill. Such transport scheduling facilitates the service of traffic agents at the mill during production. The Tabu Search and Variable Neighborhood Search metaheuristics were used to determine solutions to the model. These methodologies were able to find solutions with an average reduction of approximately 90% in vehicles' queue time, as compared with schedules produced manually by traffic agents of the mill. The solution showed also a potential savings of AU$240,000 per year compared to schedules produced manually by the traffic agent of the mill.

Mele et al. [20] formulated a multiobjective MILP model intended to optimize both the economic and the environmental performance of the production chain of cane sugar. The model is used as a quantitative tool to support decision making in the area of supply chain project planning for the combined production of sugar and ethanol, with sustainable strategic alternatives. An analysis of the model was made using a case study based on a real scenario. The authors conclude that this mathematical tool can help authorities in the analysis of strategic agroindustries and energy policies.

Silva et al. [31] proposed an integer GP model for the aggregate production planning of a Brazilian sugar and ethanol company. This model was based on conventional selection and processing techniques for the design of lots, representing the production system of sugar, ethanol, molasses, and derivatives. The work deals decisions on the agricultural and harvesting stages, sugarcane loading and transportation and energy cogeneration, selecting the production process. This approach allows decision makers to set multiple aspiration levels for their problems. An application of the proposed model for real problems in a Brazilian sugar and ethanol mill was conducted and discussed.

Ramos [26] presented a BNLP model to help solve Problem 3. The authors use strategies to calculate the model's parameters so that if the choice of harvesting dates lies outside the PUI, then the objective function suffers a penalty. This strategy also rewards harvesting dates close to the point of the cane's maximum maturing curve. Thus, the mathematical optimization model delivers an optimum plan with an estimated production figure 17.8% above production obtained by conventional means in the area in which it was applied.

Problem 4: Utilization of residual biomass to generate energy.

This problem consists in optimizing the processes involved in the exploitation of the sugarcane harvest's residual biomass for the purpose of energy generation.

Sartori et al. [29] developed a model to minimize the cost of the residual biomass transfer process, to evaluate the economics of using this material and to address sucrose production and planting area constraints, considering distances from the plots to the processing center. To solve this problem, multiobjective BLP techniques were used. The model enables one to determine an estimate of the total sucrose yield at a minimum cost and to demonstrate the economic viability of the use of the harvest's residual biomass to generate energy. Spadotto [34] proposed the application of optimization theory to improve a system designed to use the straw resulting from the mechanized harvest of sugarcane to generate energy. The goal is to maximize the volume of straw to be loaded onto the truck in the form of straw bales, thus minimizing transport costs to the processing center.

Sartori et al. [28] proposed the optimization of the sugarcane residual biomass energy balance by considering the difference between generated and consumed energy in the process of transferring this biomass from the field to the processing center. The corresponding model is a BLP model taking into account enterprise demand restrictions and cane planting area constraints. The authors concluded that using the residual biomass produced in sugarcane harvests is viable, thereby generating more energy and reducing biomass in the field. Therefore, the methodology can be applied to optimize the energy balance.

4. Conclusion

In recent years, sugarcane biomass has stood out as an alternative source of energy, both through its generation of alcohol and the cogeneration of energy through the cane's bagasse and the residue of harvesting. Thus, the growing of sugarcane has been the subject of many

http://dx.doi.org/10.5772/61533

studies from planting to harvesting to removal to the mill. These studies have the purposes of providing a management procedure at a lower economic cost and improving the quality and quantity of the cane produced. These objectives can be attained through optimized cane cultivation planning because the planning brings about several benefits, principally enhanced quality, and yield, and thus an increase in the bio-energy to be generated. However, crop planning is a complex process that requires considerable care by sector managers as it involves social, economic, political, and environmental factors. These convert the decision-making processes into issues of a multiobjective nature with significant consequences. In this case, the need arises for mathematical and computational techniques, which may assist managers in setting up the planning process for cultivation and handling of the cane. This chapter has discussed the major problems of this area which may be overcome through optimization methodologies and points to literature discussing current models and their solution methodologies.

Acknowledgements

The authors thank the Brazilian foundations CNPq (grant no. 303267/2011-9), PROPE (Pró-Reitoria de Pesquisa da UNESP), FUNDUNESP (Fundação para o Desenvolvimento da UNESP), and FAPESP (grant nos. 2014/01604-0 and 2014/04353-8) for their financial support.

Margarida Vaz Pato's research was supported by a Portuguese funding from FCT—Fundação para a Ciência e a Tecnologia, under the project UID/MAT/04561/2013.

Author details

Helenice de Oliveira Florentino[1*], Margarida Vaz Pato[3], Dylan Jones[2] and Daniela Renata Cantane[1]

*Address all correspondence to: helenice@ibb.unesp.br

1 Institute of Bioscience, UNESP, Botucatu, SP, Brazil

2 Centre for Operational Research and Logistics, University of Portsmouth, UK

3 Instituto Superior de Economia e Gestão, and CMAFCIO, Universidade de Lisboa, Portugal

References

[1] Anselmi, R. Sistematização correta reduz custo das operações no campo. Jornal da Cana, Ribeirão Preto, SP, n. 171, 36, 2008.

[2] Beeharry, R.P. Strategies for augmenting sugarcane biomass availability for power production in Mauritius. Biomass & Bioenergy, New York, v. 20, 421–429, 2001.

[3] Beeharry, R.P. Carbon balance of sugarcane bioenergy systems. Biomass & Bioenergy, New York, v. 20, 361–370, 2002.

[4] Benedini, M.S.; Conde, A.J. Sistematização de área para colheita mecanizada da cana-de-açúcar. Revista Coplana, Guariba, SP, v. 53, 23–25, 2008.

[5] Burke, E.K.; Kendall, G. Search Methodologies: Introductory Tutorials in Optimization and Decision Support Techniques. 2nd ed., 716 pages, 2014.

[6] Cherry, C. A.; Vianna, A.C.G.; Ramos, R. P.; Florentino, H. O. Problema de alocação dos talhões na área de plantio. 2014, p. 29–33. In RAMOS, R. P. Planejamento do plantio da cana-de-açúcar utilizando técnicas matemáticas de otimização. FCA Botucatu SP, 2014.

[7] Dalley, C.D.; Richard, E.P. Herbicides as ripeners for sugarcane. Weed Science, v. 58, n. 3, 329–333, 2010.

[8] Deb, K. Multi-objective Optimization Using Evolutionary Algorithms. Chichester: John Wiley & Sons, 2004.

[9] Engelke, J. Sugarcane: Measuring Commercial Quality. FARMNOTE Department of Agriculture, Kununurra, Australia, n. 23, 2002.

[10] Florentino, H.O.; Pato, M.V. A bi-objective genetic approach for the selection of sugarcane varieties to comply with environmental and economic requirements. Journal of the Operational Research Society, v. 65, 842–854, 2014.

[11] Florentino, H.O.; Lima, A.D.; Balbo, A.R.; Homem, T.P.D.; Carvalho, L.R. Multiobjective 0-1 integer programming for the use of sugarcane residual biomass in energy cogeneration. International Transactions in Operational Research, v. 18, 605–615, 2011.

[12] Florentino, H.O.; Moreno, E.V.; Sartori, M.M.P. Multiobjective optimization on economic balance from harvesting sugarcane biomass. Scientia Agricola, v. 65, 561–564, 2008.

[13] Florentino, H.O.; Sartori, M.M.P. Game theory in sugarcane crop residue and available energy optimization. Biomass & Bioenergy, v. 25, 29–34, 2003.

[14] Higgins, A. Scheduling of road vehicles in sugarcane transport: a case study at an Australian sugar mil. European Journal of Operational Research, v. 170, 987–1000, 2006.

[15] Homem, T.P.D.; Balbo, A.R.; Florentino, H.O. Optimal energy generation with biomass of sugarcane harvest. Revista IEEE América Latina, v. 2011, 653–658, 2011.

[16] Jackson, P.; McRae, T.A. Selection of sugarcane clones in small plots: effects of plot size and selection criteria. Crop Science, v. 41, n. 2, 315–322, 2001.

[17] Jones, D.; Tamiz, M. Practical Goal Programming. Springer, New York, v. 141, 2010.

[18] Lasdon, L.S. Optimization Theory for Large Systems. New York: Macmillan, 1970.

[19] Luenberger, D.G. Linear and Nonlinear Programming. Reading: Addison Wesley, 1984.

[20] Mele, F.D.; Kostin, A,M.; Guillén-Gosálbez, G.; Jiménez, L. Multiobjective model for more sustainable fuel supply chains. A case study of the sugar cane industry in Argentina. Industrial & Engineering Chemistry Research. v. 50, 4939–4958, 2011.

[21] Milan, E.L.; Fernandez, S.M.; Aragones, L.M.P. Sugar cane transportation in Cuba, a case study. European Journal of Operational Research, v. 174, 374–386, 2006.

[22] Milligan, S.B.; Balzarini, M.; Gravois, K.A.; Bischoff, K.P. Early stage sugarcane selection using different plot sizes. Crop Science, v. 47, n. 5, 1859–1865, 2007.

[23] Molin, J.P.; Portz, G.; Amaral, L.R. Precision agriculture in sugarcane production. In: Margaret Oliver; Thomas Bishop; Ben Marchant. (Org.). Precision Agriculture for Sustainability and Environmental Protection. 1st ed. Oxon: Routledge, p. 233–242, 2013.

[24] Nocedal, J.; Wright, S. Numerical Optimization. Springer. 636 p., 2000.

[25] Piewthongngam, K.; Pathumnakul S.; Setthanan, K. Application of crop growth simulation and mathematical modeling to supply chain management in the Thai sugar industry. Agricultural Systems, v. 102, 58–66, 2009.

[26] Ramos, R.P. Planejamento do plantio da cana-de-açúcar utilizando técnicas matemáticas de otimização, FCA Botucatu SP, 69 p., 2014.

[27] Ripoli, T.C.C.; Molina, F.; Ripoli, M.L.C. Energy potential of sugarcane biomass in Brazil. Scientia Agricola, v. 57, n. 4, 2000.

[28] Sartori, M.M.P.; Florentino, H.O. Energy balance optimization of sugarcane crop residual biomass. Energy, v. 32, 1745–1748, 2007.

[29] Sartori, M.M.P.; Florentino, H.O.; Basta, C.; Leão, A.L. Determination of the optimal quantity of crop residues for energy in sugarcane crop management using linear programming in variety selection and planting strategy. Energy, v. 26, n. 11, 1031–1040, 2001.

[30] Schneider, L.C.; Kinzig, A.P.; Larson, E.D.; Solórzano, L.A. Method for spatially explicit calculations of potential biomass yields and assessment of land availability for biomass energy production in Northeastern Brazil. Agriculture, Ecosystems and Environment, v. 84, 207–226, 2001.

[31] Silva, A.F.; Marins, F.A.S.; Montevechi, J.A.B. Multi-choice mixed integer goal programming optimization for real problems in a sugar and ethanol milling company. Applied Mathematical Modelling. v. 37, p. 6146–6162, 2013.

[32] Singh, K.; Gill, M.; Singh, A.; Singh, D.; Uppal, S.; Singh, J. Sugarcane planting in standing wheat using furrow irrigated raised bed (FIRB) method. Sugar Tech, v.14, n. 4, 351–356, 2012.

[33] Solomon, S. Post-harvest deterioration of sugarcane. Sugar Tech, v. 11, n. 2, 109–123, 2009.

[34] Spadotto, A.F. Teoria de empacotamento aplicada ao transporte do palhiço de cana-de-açúcar. FCA UNESP, Botucatu SP, 55 p., 2008.

[35] Stupiello, J.P. A cana-de-açúcar como matéria prima. In: PARANHOS, S. B. Cana-de açúcar cultivo e utilização. Campinas: Fundação Cargil. v. 2, 761–804, 1987.

[36] Talbe, E.G. Metaheuristics: From Design to Implementation. Wiley, 624 p., 2009.

Section 2

Biomass Uses

Chapter 3

Removal of Heavy Metals from Aqueous Solutions by Aerobic and Anaerobic Biomass

Onofre Monge-Amaya, María Teresa Certucha-Barragán,
Francisco Javier Almendariz-Tapia and Gonzalo Mauricio Figueroa-Torres

Additional information is available at the end of the chapter

http://dx.doi.org/10.5772/61330

1. Introduction

Industrialization and urbanization have promoted the generation of great quantities of aqueous effluents that may contain high levels of toxic compounds [1]. Every day, 2,000,000 tons of wastes (from sewers or agricultural and industrial residues) are released into rivers and seas, spreading disease and damage to ecosystems. Achim Steiner, executive chief of the United Nations Program for the Environment stated: "If the world is to thrive, let alone to survive on a planet of 6 billion people heading to over 9 billion by 2050, we need to be collectively smarter about how we manage waste, including wastewaters" [2].

Heavy metals, or potentially toxic elements, constitute a specific group of pollutants that are released into the environment as a result of industrial activities, such as the mining industry. These elements can cause health problems. In México, the mining industry is one of the most important economic activities, with gold, silver, and copper being the precious metals with higher production rates [3].

The metallurgical process of the mining industry involves a series of extraction and purification techniques that result in the disposal of metals into water bodies through acid mine drainage (DAM). Heavy metals can then accumulate at toxic concentrations for a functional ecosystem, which constitutes an economic problem of public health [4].

Controlling and reducing water pollution is a significant concern for our society. Wastewater spills create eutrophication and toxic problems. The wastewater penetrates the soil, contaminates groundwater, and reduces the quality necessary for human consumption [5].

Discharge limits have been established for heavy metals, among many other water pollutants. Most heavy metals are soluble and form aqueous solutions; hence, they cannot be separated by ordinary physical treatments [6].

Contamination of soil and water is the result of numerous industrial activities such as mining, melting, fabrication of jewelry, batteries, and automobiles, and volatile ashes from incineration processes. This type of contamination poses a serious threat for human and animal health since heavy metals remain in the environment for an indefinite time [7].

México has several sites contaminated by heavy metals and other residues from the mining industry. A particular example of pollution is found in the San Pedro River, located in the state of Sonora, México, where silver and copper production has been exploited for decades. The San Pedro River stream originates near Cananea – a mining town known for having the biggest mining districts of the state – and culminates in the state of Arizona, in the United States. Surface water pollution in the San Pedro river was reported in 1997 and 1999 [8]. In 2008, the presence of heavy metals in the river sediments was also evaluated [9]. The river has been contaminated by heavy metals due to its proximity with the metallurgical activity of the state. Metals found in the river are: cadmium, cobalt, chromium, iron, manganese, copper, zinc, nickel, and lead. However, two of the metals with higher concentrations were copper and iron, which exceeded the maximum permissible values established in the Mexican laws for water quality. These laws consider lead, zinc, mercury, silver, nickel, cadmium, aluminum, copper, and arsenic, as water pollutants due to the toxicity they pose for aquatic and terrestrial organisms (NOM-001-ECOL-1996; NOM-002-ECOL-1996; NOM-003-ECOL-1996).

More recently, on August 7, 2014, the Buenavista Copper Mine in Cananea was under the spotlight when approximately 40,000 cubic meters of sulfuric acid were spilled into the Bacanuchi River (also situated in Sonora). This toxic leakage affected an estimate of about 800,000 people [10]. Heavy metals pollution has been reported, but the remediation projects aiming to recover the quality of these sites have been extremely scarce. Thus, it is of great importance for research institutions and industries to evaluate technological alternatives for the removal and stabilization of inorganic contaminants, keeping into consideration the specific environmental conditions of each polluted site [11].

2. Heavy metal removal processes

The removal of heavy metals can be carried out by a number of conventional treatments, such as reverse osmosis, electrodialysis, ultrafiltration, chemical precipitation, and ionic exchange. These methods, however, have the disadvantage of requiring high operation costs. The ionic exchange resins, for example, have been commercially known for their effectiveness as pollutant adsorbents in wastewater treatments, but their high cost hinders their application at industrial levels [1]. Chemical processes, although simple to perform, end up being even more expensive because of the active agent that cannot be recovered for future uses. Besides, the final product is a high concentrated sludge difficult to handle [4].

Heavy metals sources are not renewable, and the natural reserves are being consumed. Therefore, it is imperative that those elements considered dangerous to the environment or those of technological importance and economic value are withdrawn and recovered at their point of origin through appropriate treatments.

A brief description of the before mentioned processes is presented next:

Reverse osmosis: a process where heavy metals are separated through a semipermeable membrane by using a pressure higher than the osmotic pressure, which is caused by the dissolved solids in wastewaters. The high pressures required for this process are the main reason for the high operating costs of reverse osmosis.

Electrodialysis: in this process, metallic ions are separated by selective semipermeable membranes. An electric current is applied between two electrodes located at each side of the membranes, which produces a migration of cations and anions toward their respective electrodes. The migration of ions results in the formation of metal salts that precipitate out of solution. However, a major disadvantage of electrodialysis is membrane clogging, caused mainly by the formation of metal hydroxides.

Ultrafiltration: this process involves the use of porous membranes and high pressures for the separation of metal ions. Sludge generation is the main disadvantage in this treatment.

Ionic exchange: metallic ions in diluted solutions are exchanged with the ions located in the active sites of synthetic resins by electrostatic forces. Sludge generation and the high costs of exchange resins are the main disadvantages.

Chemical precipitation: precipitation of metallic ions is achieved by the addition of coagulants such as calcium salts, iron, and other organic polymers. The inconvenience of this method is the excessive amounts of sludge (it might include toxic compounds) produced during the precipitation.

Phytoremediation: it involves the use of certain plants as removing or stabilizing agents in contaminated soils, sediments, and water. The time required for effective stabilization of heavy metals is large and can be a constraint in this process; furthermore, plant regeneration is even more complex.

All of the disadvantages previously mentioned, such as incomplete removal, high energy consumption, excessive residual sludge, and formation of other toxic residues requiring careful disposal protocols justify the need for a cost-effective treatment for the removal of heavy metals from wastewater [12].

New technologies are currently being developed, taking into consideration the processing costs and direct scaling up and implementation [13]. The search for effective removal technologies has directed attention toward biosorption, an ecological alternative that uses different biological materials for binding and concentrating metal ions.

Biosorption: This process is based on the capacity of biological materials to concentrate heavy metals by either metabolic or physical–chemical pathways.

Developments in the field of environmental biotechnology have allowed the identification of several species of algae, bacteria, fungi, and yeast as effective metal biosorbents [14]. The main advantages of biosorption over conventional treatments include: lower costs, high removal yields, minimum residual sludge formation, and potential biosorbent regeneration and metal recovery [15].

The biosorption process involves a solid phase – the biosorbent, or biomass – and a liquid phase – the solvent (commonly water). The liquid phase contains the sorbate, i.e., the species to be sorbed (metallic ions). During biosorption, the sorbate is attracted and bound to the biosorbent through a variety of mechanisms. This "binding" process continues until a state of equilibrium is achieved between the amount of sorbate present and the available active sites of the biosorbent [16].

The two mechanisms by which biosorption can take place are [13]:

- Bioaccumulation: based in the intracellular transport of metallic ions by living biomass.
- Bioadsorption: based on the adsorption of metallic ions on the cell surface. This process can occur by ionic exchange, precipitation, complexation, or electrostatic attraction. Figure 1 shows a basic experimental approach that can be used to determine the biosorption capacity, q, a measure of the metal uptake by biomass.

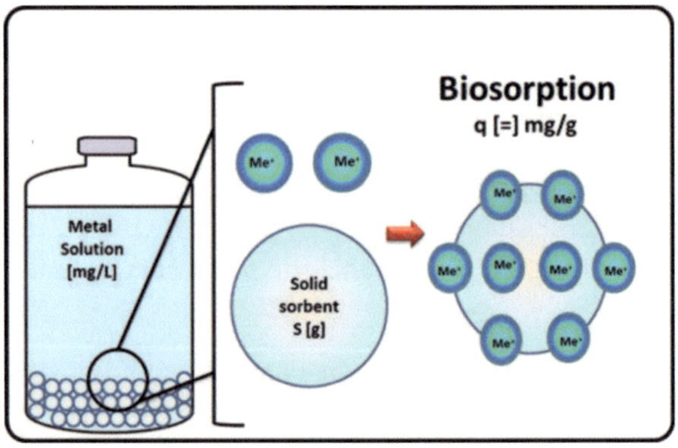

Figure 1. General experimental setup for biosorption of heavy metals.

The biosorption process can be carried out in a bioreactor, where the wastewater flows through a bed of microorganisms which bind the heavy metals. Bioreactors are useful tools where high volumes of wastewaters may be treated continuously, transferring the contaminated "portion" to a considerable smaller volume. However, certain problems can arise during the operation of bioreactors, such as biomass washout, liquid–solid separation difficulties, and pressure drops. These problems originate due to the fact that microbial biomass generally consists of small particles with low density and poor mechanical strength [17]. Immobilization of biomass in a suitable matrix (or material supports) can overcome washout problems by inducing cellular growth in the form of a stable biofilm constituted by microbial cells and extracellular polymeric substances.

3. Heavy metals removal by aerobic biomass

Nowadays, the use of microorganisms for environmental remediation and recovery purposes has grown as a research field. It is believed that the most fitted microorganisms for removal treatments are the ones isolated from the same environment where they were naturally selected; however, genetic manipulation techniques can be used to enhance the capacity of different microorganisms [18].

Bioremediation utilizes the catalytic abilities of living organisms to degrade and transform pollutants from aquatic and terrestrial ecosystems. This alternative can be potentially applied to mitigate environmental contamination. Bioremediation has focused on the exploitation of genetic diversity and metabolic versatility, characteristic traits that make bacteria suitable for the transformation of pollutants into harmless products, or less toxic compounds, that can be reintegrated in the natural biochemical cycles. On the other hand, there are other microorganisms such as fungi or plants that have been isolated and used in removal processes like phytoremediation [19].

Microorganisms are naturally exposed to heavy metals in essential or toxic quantities, and the amount of heavy metals in certain sites can be so high that microorganism growth is not possible. Metal toxicity forces microorganisms to develop various strategies to defend themselves against high concentrations of heavy metals [20].

There are several experimental protocols important to effectively examine metal biosorption by aerobic biomass. These protocols are described below.

3.1. Isolation / Inoculation

Isolation is used to identify microorganisms able to grow in polluted environments. Wastewater samples are generally collected from damaged sites, and yeast or bacteria (biomass) cells are grown by inoculating them into a nutrient-rich environment. Inoculation is usually done in cell-culture dishes by the streaking method using selective enriched nutritive media for each microorganism. Commonly, 10 mL of wastewater sample is inoculated in a specific culture medium at 37°C for 24 h.

3.2. Batch biosorption and kinetics of heavy metals

The biosorption batch tests with aerobic biomass are carried out in experimental vessels, such as Erlenmeyer flasks. Wastewater samples are mixed with a known amount of biomass. Flasks are placed in an incubator at specific conditions and tests are carried out in duplicate, using two flasks for every sampling time. For aerobic microorganisms such as yeast, the conditions are usually set as follows: pH 3–4, 37°C, and 100 rpm. Samples are taken at regular intervals until equilibrium is achieved. Every sample is then centrifuged to separate biomass from the solution. Concentration of metals is usually determined by atomic absorption spectrometry.

Biosorption efficiency (E) is calculated as follows:

$$E = \left(\frac{Co - Cf}{Co} \right) \times 100 \tag{1}$$

where:

C_o, C_f are the initial and final metal concentration (mg/L).

The biosorption capacity of the biomass at any given time is calculated as follows:

$$q_{eq} = \frac{m_o - m_{eq}}{V_{ads}} \tag{2}$$

where

m_o is the initial mass (mg), equal to the initial concentration (mg/L) times initial volume;

m_{eq} is the mass at equilibrium (mg), equal to the concentration (mg/L) times volume at equilibrium;

v_{ads} is the volume of biomass used (L).

3.3. Continuous biosorption studies

Continuous studies are carried out in bioreactors. Bioreactors consist commonly of acrylic or glass columns with lateral sampling points. Perhaps, the simplest configuration is the Upflow Aerobic Reactor packed with material supports and biomass recirculation. An example of material support is clinoptilolite, a zeolite with a particle size of 4.76 mm, a pore diameter of $3.22E^{-03}\mu$ m and a Si/Al ratio of 4.53. Aerobic conditions are met by supplying air from the bottom of the column through peristaltic pumps.

3.4. Biosorption tests in aerobic reactors

Once reactors are inoculated with the selected aerobic biomass, mineral medium is used for biomass acclimation at pH levels optimum for growth. Mineral medium is only used during startup as a source of nutrients for biomass growth and immobilization. In the case of yeasts, pH is generally 3–4, and the medium consists of the following compounds: (g/L): ammonium phosphate 1, glucose 5, sodium chloride 5, magnesium sulfate 0.2, and phosphate potassium 1 [21].

Figure 2 shows a schematic diagram of two Upflow Aerobic Reactors connected in series that were used to remove heavy metals by Hernández-Mata et al., 2014 [22]. In this scheme, the first reactor (R1) was inoculated with biomass and the effluent was recirculated until the biomass reached a concentration of 1 g/L. When the desired biomass concentration was achieved, the biosorption stage was initiated with mining effluents. After the biosorption stage, a desorption (purification) step was carried out to remove the metallic ions adsorbed by the biomass. Biomass concentration was measured once again until the concentration reached 1 g/L. The effluent of R1 was then fed to R2 (containing the same biomass produced in R1) and biosorption was examined in both reactors. Samples were taken at regular intervals at the inlet and outlet points until column saturation was evident [22].

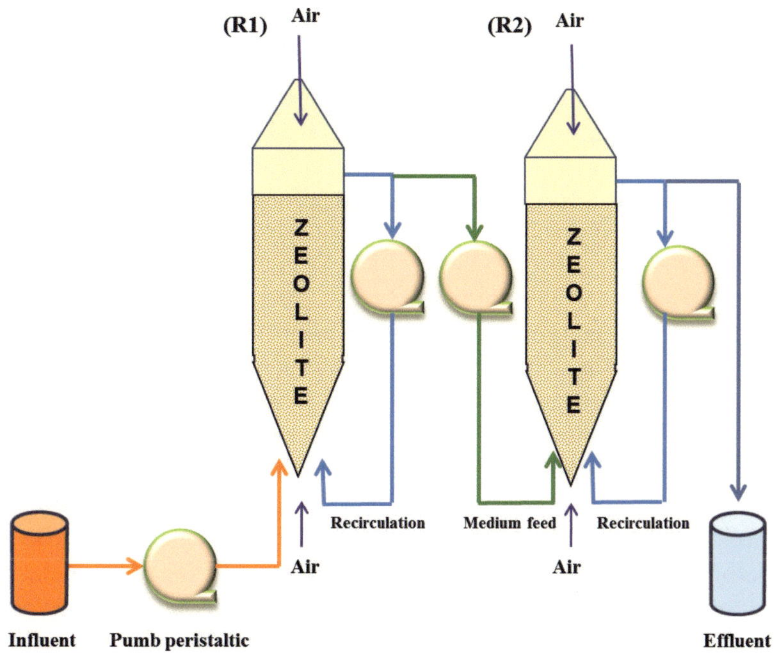

Figure 2. Schematic diagram of two upflow aerobic reactors packed with zeolite.

4. Heavy metals removal by living anaerobic biomass

Anaerobic microorganisms perform as part of their metabolism a process known as anaerobic digestion, which has been widely implemented in the treatment and stabilization of effluents with high organic loads. Two of the main bacterial groups that participate in anaerobic digestion are acidogenic microorganisms (responsible for the conversion of organic matter into volatile fatty acids, VFAs) and methanogenic microorganisms (methane producers).

Generally, it is considered that methanogenic bacteria are less resistant to external changes in their growing conditions such as pH, temperature, and/or presence of toxic metals [23]. It was also reported in a previous study that inhibition by heavy metals was less noticeable for acidogenic bacteria [24].

4.1. Biomass treatment (Acidogenic phase)

To achieve acidogenic conditions, biomass can be inoculated in Erlenmeyer flasks for a large period of time (up to 8 weeks), mixing anaerobic sludge and material supports (if desirable). The flasks are kept at 30°C. The feed medium is changed continuously and prepared according to the requirements of the microorganism [25]. The medium pH is kept at acidic levels (3–4) to inhibit the growth of methanogenic organisms, which is favored at neutral pH.

Dextrose is generally used as substrate. This substrate is the source of organic matter that enhances volatile fatty acids (VFAs) formation, mainly: acetic acid, propionic acid, and butyric acid. In order to verify that the anaerobic sludge is carrying out the acidogenic phase of digestion, VFAs formation and concentration can be measured by HPLC (high performance liquid chromatography) taking samples from the flasks at regular intervals. pH can be measured daily and the growth of biomass can be indirectly calculated by determining the volatile suspended solids (VSS), which are obtained according to the gravimetric method [26].

4.2. Toxicity studies

Toxicity tests are carried out prior to any biosorption test with living biomass to obtain inhibitory concentrations. For acidogenic biomass, VFAs formation or substrate consumption are a direct measurement of microbial activity. During a toxicity experiment, a known amount of biomass (or immobilized biomass, if desirable) is put into a series of flasks and mixed with fixed volumes of metallic solutions and a selected substrate. The concentration of heavy metals in the metallic solutions varies according to each experimental setup and metallic ion, but one flask must be selected as a blank. The concentration of the organic substrate is kept constant in all flasks. Solution pH has to be adjusted to acidic levels (3–5) to avoid metal precipitation. Once the biomass and solutions are mixed, the flasks are closed and placed in an incubator at a specific temperature and rpm (for instance, 35°C and 50 rpm). Small liquid samples are taken from each flask at regular intervals to determine substrate or VFAs concentration. Sampling can stop when concentrations in all flasks remain constant for at least two consecutive points.

Once all measurements are done, toxicity is determined in terms of the half-inhibitory concentration, IC_{50}, which is the concentration at which microbial activity is decreased by 50%. Microbial activity is determined by calculating the difference between the initial concentrations and final concentrations in each flask and dividing it by the concentration difference of the blank (Equation 3). IC_{50} is then determined graphically from a plot of "% activity" versus "metallic concentration". The blank is considered to have a 100% microbial activity since no metallic inhibition takes place, but the activity decreases with increasing heavy metal concentration.

$$A(\%) = \frac{C_{0-}C}{C_{B0-}C_{Bt}} \times 100\% \tag{3}$$

where

A(%): microbial activity.

D_0, D_{48}: concentration of substrate or VFAs at times 0, and t, respectively. D_{c0}, D_{c48}: concentration of substrate or VFAs in the blank flask at times 0, and 7, respectively.

4.3. Biosorption isotherms

Biosorption isotherms are plots of biosorption capacity versus metallic concentration at equilibrium. Isotherms can be adjusted to adsorption models to determine other parameters

http://dx.doi.org/10.5772/61330

useful in the scaling up of biosorption processes, such as maximum biosorption capacity and affinity coefficients. To determine biosorption capacity, batch tests are carried out in a similar fashion to toxicity tests, but the variable of importance is the heavy metals concentration. A known amount of biomass (or immobilized biomass, if desirable) is put into a series of flasks and mixed with fixed volumes of metallic solutions. Metallic ions concentrations are determined by atomic absorption spectrometry. Biosorption equilibrium takes place when concentrations in all flasks remain constant for at least two consecutive points, and sampling can stop. Biosorption capacity can then be calculated according to Equation 4 [27].

$$q = \frac{V(C_0 - C_f)}{S} \tag{4}$$

where

q = biosorption capacity, (mg metal/g VSS);

C_0 = initial metal concentration (mg metal/L);

C_f = final metal concentration (mg metal/L);

S = biosorbent (biomass) used (g);

V = volume of metallic solution (L).

The data at equilibrium (concentration and biosorption capacity) can be adjusted to established adsorption models. A correlation factor can be calculated by lineal regression to determine which model fits best to the experimental values. The most commonly used models in the literature are the Langmuir and Freundlich models.

4.4. Continuous studies

Continuous studies can be carried out in bioreactors of all shapes and sizes, but the most commonly used configuration is the anaerobic packed bed reactor (APBR). Generally, wastewater flows upward through the reactor bed, and the use of a material support prevents from biomass losses and enhances bed stability. Environmental conditions depend upon the type of biomass used. Figure 3 shows the schematic diagram of an APBR used for the biosorption of heavy metals [28]. Bioreactors startup times are varied, and the parameters commonly measured during operation are pH, chemical oxygen demand (COD), substrate consumption, methane formation, VFAs formation, volatile suspended solids (VSS). Recirculation of the effluent can be added to the reactors configuration to enhance biomass growth before biosorption takes place.

The COD values are a measure of the organic load of wastewaters. When both the influent and effluent points are sampled, the COD analysis provides a quantifiable measurement of the removal efficiency of organic matter in the bioreactor. The most common COD method involves digestion of the sample at 120°C followed by a colorimetric analysis. The procedure is thoroughly described in the Standard Methods for the Examination of Water and Wastewater [26].

VFAs concentrations are indicative of the acidogenic activity of anaerobic biomass. Total VFAs can be analyzed by a simple titration method (using hydrochloric acid and sodium hydroxide) proposed by [29] Powell and Archer (1989). Specific VFAs, such as acetic acid, propionic acid, or butyric acid can be analyzed by HPLC. For the determination of substrate consumption, most methods are relatively simple and involve colorimetric techniques. A method utilized for glucose concentration is the DNS (3,5-dinitro-salicyclic acid) method, where the free sugar reduces the DNS reagent at high temperature, resulting in the formation of a colored product that absorbs light at 540 nm [30].

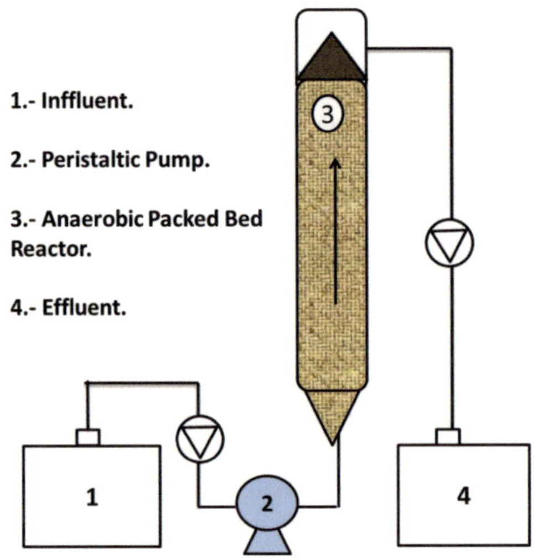

Figure 3. Example of an APBR used for heavy metals biosorption.

Once the startup stage is complete, heavy metals can be fed to the bioreactor to initiate the biosorption stage. A plot of C/C_0 versus time is known as a rupture curve, where C_o is the inlet concentration and C is the outlet concentration. Rupture curves provide information about the quality of a biosorbent in terms of the breakthrough time, saturation time, and retention capacities. The breakthrough time, t_b, is defined as the time in which the outlet concentration is equal to a maximum permissible value (usually 10% of the inlet concentration or lower). Saturation time, t_s, is the time in which the column is completely saturated by the metallic ions. Metallic retention capacity, Q_{ads}, can be calculated according to the following equation:

$$Q_{ads} = \frac{C_0 F}{m_s} \int_{t=t_0}^{t=t_s} \left(1 - \frac{C}{C_o} \right) dt \quad \Rightarrow \quad Q_{ads} = \frac{F}{m_s} \int_{t=t_0}^{t=t_s} C_{ads} dt \qquad (5)$$

where

Q_{ads}: Retention capacity [mg/gVSS];

C_{ads}: C_o-C [mg/L];

t_0: Initial time [d];

t_s: Saturation time [d];

F: Volumetric flow [L/d].

Removal efficiency can also be determined simply by calculating the total metallic load and final metallic retention.

4.5. Bed characterization

Bed characterization in anaerobic reactors is usually achieved by the following techniques: fraction of solids, Gram staining, microscopic observation via optical microscopy or scanning electron microscopy (SEM), X-ray diffraction (XRD), and energy dispersive spectroscopy (EDS). These analyses supply plenty of information about the morphology and structure of the microorganisms and extracellular polymeric substances of the biofilm. XRD and EDS are especially helpful when a material support is utilized since these analyses provide the elemental composition of the different solid phases of the bioreactor bed.

5. Sulfate-reducing process and metal bioprecipitation

The microbial sulfate-reducing process (SRP) has been utilized as a potential tool for heavy metals removal during the final steps of wastewater treatments and effluent recovery of several industries. Under anaerobic conditions, sulfate-reducing bacteria (SRB) reduce sulfate to sulfur, which reacts with the metallic ions, resulting in the formation of metallic sulfurs. Metallic sulfurs are universally identified because of their low solubility in aqueous systems, making the sulfate-reducing process an effective alternative for wastewater treatment [31]. Furthermore, selective recovery of economically important metals is also possible [32]. The sulfate-reducing process is successfully applied in the removal of metallic ions and sulfates in acid mine drainage (AMD), and can be useful in the removal of the remaining metals in industrial wastewaters [31].

During the SRP, sulfate ions $\left(SO_4^{2-}\right)$ are enzymatically reduced to sulfur (H_2S, HS-, and S^{2-}) to obtain the energy required for the growth and maintenance of SRB. In order for this process to take place, cells carry out an enzymatic oxidation of organic matter (electron donor) to carbon dioxide and water [33, 34]. The SRP is strictly an anaerobic process, since it can only occur in the absence of electron acceptors with high redox potential such as oxygen or nitrate [35]. The SRP for heavy metals removal is based on the formation of metallic sulfurs with low solubility and the neutralization of water as a result of the alkalinity produced during the microbial oxidation of the electron donors [36]. This phenomenon has been defined as bioprecipitation [37], and can be described by the following equations [38].

Formation of sulfur and alkalinity (sulfidogenic oxidation) is defined by Equation 6, where CH_2O represents the electron donor:

$$2\ CH_2O + SO_4^{2-} \rightarrow H_2S + 2\ HCO_3^- \tag{6}$$

When H_2 is used as electron donor, the reaction generates hydroxide ions:

$$8H_2 + 2SO_4^{2-} \rightarrow H_2S + HS^- + 5H_2O + 3OH^- \tag{7}$$

The formation of biogenic sulfur (H_2S, HS-, S_2-) enhances precipitation of dissolved metals, where M2+ represents metallic ions such as: Zn2+, Cu2+, Ni2+, Co2+, Fe2+, Hg2+, Pb2+, Cd2+, o Ag+:

$$H_2S + M^{2+} \rightarrow MS_{(s)} + 2\ H^+ \tag{8}$$

The precipitation of metallic ions releases protons which acidify the water. Consequently, it is necessary to reduce the excess of sulfate to compensate acidity. The alkalinity of the hydroxide ions or bicarbonate produced during the sulfidogenic oxidation neutralizes the acidity of water.

$$HCO_3^- + H^+ \rightarrow CO_{2(g)} + H_2O \tag{9}$$

$$OH^- + H^+ \rightarrow H_2O \tag{10}$$

5.1. Advantages of the sulfate reducing process in wastewater treatment

The SRP is a valuable biotechnological tool for heavy metals removal in mining lixiviates and industrial effluents. It is considered potentially superior to other biological processes due to its capacity to produce alkalinity, neutralize the pH of acidic water, and simultaneously remove organic matter, sulfates and heavy metals [39, 32, 40, and 38]. Furthermore, recent studies of the SRP have revealed potential immobilization for metalloids (arsenic), radioactive isotopes (uranium), and cyanides [41, 42, and 43]. The SRP has also shown applications in organic matter removal and degradation of xenobiotic and toxic compounds [44].

The most commonly known advantages of the SRP are the low formation of metallic sulfur sludge (small volume and low solubility) compared to hydroxide precipitation and the recovery of economically important metals and precipitated metallic ions [45]. Recently, some methods have been implemented to selectively recover metals through pH and sulfur control [33].

http://dx.doi.org/10.5772/61330

5.2. Toxicity of metals

It has been reported that metals are inhibitory agents for anaerobic microorganisms, including SRB [46, 47]. The inhibition is mostly due to the capacity of metals to deactivate enzymes by reacting with other sulfhydryl groups (-SH) and replacing the metals that constitute the active sites, such as Cu(II), Zn(II), Co(II), Ni(II). The deactivation of enzymes implies a negative impact on bacterial growth and activity [48]. There are some discrepancies in the literature with regard to the inhibitory levels of heavy metals over SRB because the majority of experiments are carried out at different environmental conditions [49].

Biogenic sulfur (produced during the SRP) forms complexes insoluble with heavy metals, resulting in the precipitation of metallic sulfur and, in turn, a toxicity reduction [46]. Sulfur inhibition may be decreased by precipitating sulfur with iron [50]. Several studies have focused on the use of SRP for the precipitation of metallic sulfurs within the same reactors where the sulfate-reducing activity takes place. However, this method might increase the inhibition of SRB [51].

To reduce inhibitory effects and increase pH in anaerobic reactors, a portion of the wastewater can be recycled and mixed with the influent. The remaining sulfur in the recirculating effluent reacts with heavy metals and causes precipitation of metallic ions before they get in contact with the anaerobic sludge [52]. The search of new strains tolerant to sulfurs or the special designs of bioreactors can help to prevent the toxic effect of heavy metals on SRB [53].

Another problem associated with heavy metal precipitation within the reactor is that metallic sulfurs are deposited on the biomass, and the contaminated sediments generate an increase in volume [54]. Moreover, contrary to general belief that only soluble metallic ions cause inhibition, it has been proven that metallic sulfurs affect the metabolic activity of SRB. Metallic sulfurs are not toxic, but they block substrate and nutrients access into the cells by forming a barrier on the cellular walls of SRB [47]. A proper alternative to separate the biological process from the precipitation is to use a two-step process, where metallic precipitation is isolated from the biological process [54].

5.3. Selective precipitation of heavy metals

Metallic sulfurs are generally highly insoluble at neutral pH, whereas some compounds, such as CuS, are insoluble at pH values as low as 2. The great advantage of precipitation is the possibility for selective recovery of metallic sulfurs. It has been shown that each metal precipitates at a unique sulfur concentration S^{2-}, or potential (pS), directly related to the solubility of the metallic sulfur formed. Controlling these concentrations within a precipitator can be carried out using pH electrodes and sulfide ion selective electrodes (pS electrode). The unique quality of the potential level (pS) of each metal has been successfully applied as a controlling parameter for the selective precipitation of metals and formation of pure metallic sulfurs suitable for reutilization. The success of the precipitation process depends not only on the heavy metal removal from the soluble phase but also on its separation from the liquid phase. Thus, solid–liquid separation processes (for instance, sedimentation and filtration) are of great importance for a successful removal [55].

5.4. Types of reactors used for the sulfate-reducing process

Biomass is retained within bioreactors according to the adherence properties of cells. Thus, bioreactors can be classified into two groups [56]: fluidized bed reactors and fixed bed reactors. In a fixed bed reactor, biomass is retained either by the formation of biofilms on static or suspended inert materials or by the obstruction of biological particles on packing materials (Figure 4). A biofilm is defined as a complex structure constituted by cells and extracellular products in elongated or granular forms [57]. In fluidized bed reactors (or free bed reactors), biomass is retained by forming biological particles of high density and sedimentability: granules. Methanogenic granular sludge and sulfate-reducing sludge are composed of microbial aggregates that grow by mutual bonding of bacterial cells in the absence of a support material [58].

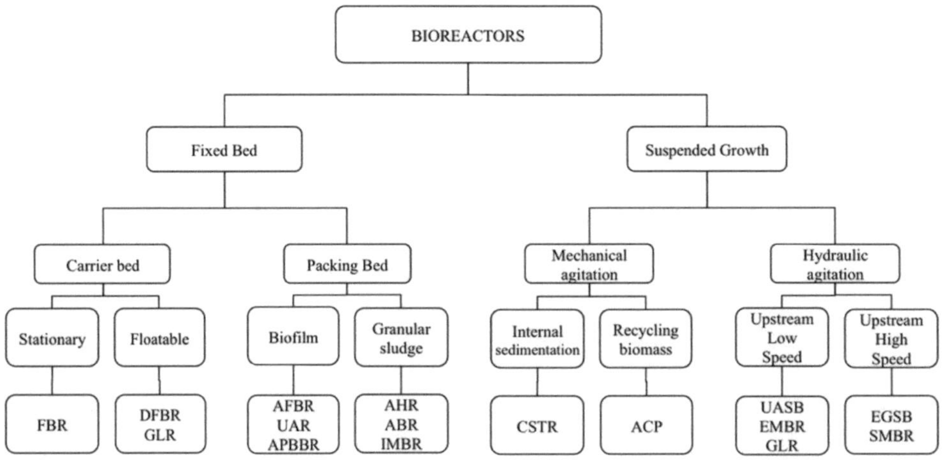

Figure 4. Anaerobic reactors used in sulfate-reducing applications.

Numerous literature studies have applied multiple reactors designs of the sulfate-reducing process for the treatment of water with high concentrations of sulfates and heavy metals. Some of these designs include batch reactors (BR), sequencing batch reactors (SBR), continuously stirred tank reactors (CSTR), anaerobic contact processes (ACP), anaerobic baffled reactors (ABR), anaerobic filter reactors (AFR), fluidized bed reactors (FBR), gas lift reactors (GLR), anaerobic hybrid reactors (AHR), membrane bioreactors (MBR), and upflow anaerobic sludge blanket reactors (UASB) [38].

6. Biosorption models

The kinetic model of a microbial process is defined as: the verbal or mathematical correlation between velocities and concentrations of reagents products, inserted into mass balances for the prediction of substrate conversion level and individual yields at specific operating conditions [59].

The complexity of the kinetic models used to describe the changes within the cell during a microbial transformation can be very broad. Several kinetic models proposed in the literature are summarized in Table 1 [60].

Kinetic model	Definition
Unstructured – non segregated	Biomass is considered the only component. An average cell is representative of the microbial consortium.
Metabolic	Metabolic pathways are described as a network of reactions using a simplified reaction scheme. Stoichiometric relations are defined.
Sructured (or Cell)	Biomass is considered to be constituted by several species. Intracellular components are taken into account.
Segregated	The distribution of a property is considered in the description of the biomass.

Table 1. Classification of kinetic models.

The simplest models, unstructured-non segregated models, have been used for numerous engineering troubleshooting applications. However, in order to have a better system description it is necessary to use models that take into account complex reaction schemes, i.e., models that take into account the metabolic pathways of each microorganism.

A mathematical model is the abstract representation of a specific aspect of reality. Its structure is composed of two parts. The first part corresponds to all those characteristic aspects of an idealized reality, and the second part refers simply to the existing relationships between the aforementioned elements [61].

The order of reaction is an experimental magnitude dependent of the way in which velocity relates to concentration [62]. Any typical reaction in nature will occur at a rate dependent of certain factors, the reaction rate is indicated by a constant value (k). It is found that reaction rates are related to the reaction order according to the following mathematical expression [63]:

$$-\frac{dA}{dt}=k\left[\mathrm{A}\right]^{\mathrm{n}} \tag{11}$$

where

n = reaction order;

k = rate constant;

A = concentration of component A;

t = time.

This equation is integrated for every order of reaction (zero order, first order, second order, pseudo first order, and pseudo second order) as follows:

- *Zero order reaction:*

Differential equation:

$$-\frac{dA}{dt} = kA^{\circ} \quad (12)$$

Separating variables:

$$\int dA = k\int dt \quad (13)$$

Solving the integral:

$$A = kt + C \quad (14)$$

- *First-order reaction:*

Differential equation:

$$-\frac{dA}{dt} = kA^{1} \quad (15)$$

Separating variables:

$$\int \frac{dA}{A} = k\int dt \quad (16)$$

Solving the integral:

$$LnA = kt + C \quad (17)$$

- *Second-order reaction:*

Differential equation:

$$-\frac{dA}{dt} = kA^{2} \quad (18)$$

Separating variables:

$$\int \frac{dA}{A^{2}} = k\int dt \quad (19)$$

Solving the integral:

$$\frac{1}{A} = kt + C \tag{20}$$

where C: Integration constant

- *Pseudo first-order reaction:*

Differential equation:

$$\frac{dx}{dt} = k_2(C_{A0} - x)b \tag{21}$$

Solving the equation:

$$k = bk_2 = \frac{1}{t}\left[\frac{1}{C_A} - \frac{1}{C_{A0}}\right] \tag{22}$$

- *Pseudo second-order reaction:*

$$\frac{dx}{dt} = k_2(C_{A0} - x)^2 b \tag{23}$$

Solving the equation:

$$k = bk_2 = \frac{1}{t}\left[\frac{1}{C_A} - \frac{1}{C_{A0}}\right]^2 \tag{24}$$

where

C_A = amount of metal adsorbed (mg/L)

C_{Ao} = initial concentration (mg/L)

t = time (min)

k = equation constant (mg/L-min)

b = initial concentration of component b, constant throughout the reaction time.

If the lineal model properly fits the experimental values (i.e., a correlation factor, R^2, close to 1) the adsorption process can be described as chemisorption [64].

The development of biosorption systems is dependent of many factors including: temperature, pH, biosorption capacities and selectivities, recovery efficiency, and resistance to other components or operating conditions. Nevertheless, most biosorption studies focus on the measurement of the biosorption capacities of biomass [65]. The quantification of the sorbate–biosorbent interactions is fundamental for the evaluation of the biosorption capacity. Due to the similarity between the biosorption process and the adsorption process, biosorption capacity can be analyzed by sorption isotherms. Sorption isotherms are model equations that represent the behavior of experimental data.

The Langmuir and Freundlich equations are two of the most utilized adsorption models. These models are described by the following equations [1]:

6.1. Langmuir model

$$q_e = \frac{q_{\max} b C_e}{1 + bC_e} \tag{25}$$

$$\frac{1}{q_e} = \left(\frac{1}{bq_{\max}}\right)\left(\frac{1}{C_e}\right) + \left(\frac{1}{q_{\max}}\right) \tag{26}$$

where

q_e: biosorption capacity at equlibrium (mg/g VSS).

q_{max}: máximum biosorption capacity (mg/g VSS).

C_e: metallic concentration at equilibrium (mg/L).

b: affinity coefficient between the sorbate and the biosorbent (L/mg).

q_e and C_e are obtained at the equilibrium point, whereas q_{max} and b can be determined graphically by a plot of $(1/q_e)$ versus $(1/C_e)$.

6.2. Freundlich model

$$q_e = kC_e^{1/n} \tag{27}$$

$$\mathrm{Ln}(q_e) = \frac{1}{n}\mathrm{Ln}(C_e) + \mathrm{Ln}(k) \tag{28}$$

where

q_e: biosorption capacity at equilibrium;

C_e: metallic concentration at equilibrium;

k,n: Freundlich constants.

The parameters k and n can be graphically determined from a plot of $\mathrm{Ln}(q_e)$ versus $\mathrm{Ln}(C_e)$.

7. Conclusion

Environmental pollution is one of the main problems of our society. Heavy metals constitute a major group of contaminants characterized by having a density five times greater than that of water. One of the main sources of heavy metals pollution is the acid mine drainage (AMD) generated by mining industries. The AMD is an acid lixiviate that may contain high concentrations of sulfates, iron, calcium, zinc, manganese, aluminum, copper, and other types of toxic elements such as arsenic and lead. In México, several regions have been affected due to the presence of heavy metals in wastewaters, which generates the necessity of implementing economic and efficient remediation techniques. The review focuses on biological methods and the advantages they offer over conventional treatments. One particular alternative studied in recent years is biosorption – based on the ability of biomass to bind and concentrate heavy metals – because of its economic nature and high removal efficiencies in dilute wastewaters.

Biological technologies provide plenty of advantages and can be just as effective and economic as other technologies (Table 2). However, it is of upmost importance to continue with scientific research to acquire an improved understanding of the bioremediation processes and optimize industrial applications.

Method	Advantages	Disadvantages
Chemical precipitation	-Low cost -Simple operation	-Excessive formation of sludge -Slow or insufficient precipitation
Reduction	-No residual sludge generation	-Formation of toxic gaseous products -Difficult handling of reagents
Ultrafiltration	-Requires small operating space -High separation	-High operating costs -Membrane clogging
Reverse osmosis	-High efficiency -Removal of other ionic compounds	-High operating and maintenance costs
Electrodialysis	-High efficiency	- High energy requirements (high pressures) -Clogging
Adsorption-Ionic Exchange	-Simple operation -High adsorption capacity	-Low selectivities - pH sensitive
Phytoremediation	-Eco-friendly technology and low cost -Removes other types of pollutants	-Depends on the growing conditions. Long remediation times. It requires expensive agricultural equipment.
Microbial bioremediation	-Eco-friendly technology and low cost -Generates no toxic waste (CO_2, H_2O)	-Limited pollutants range -Microbes need proper growing conditions.
Biosorption	-Low cost -Minimum formation of residual sludge -Potential for metal recovery -Simple operation -Effective in diluted solutions.	-Toxic effects on living biomass. -Constant nutrient supply for biomass growth.

Table 2. A comparison between the existing methods for heavy metals removal.

Acknowledgements

The authors would like to thank the Department of Chemical Engineering and Metallurgy of the University of Sonora for the space and equipment facilitated during the realization of this work.

Author details

Onofre Monge-Amaya*, María Teresa Certucha-Barragán,
Francisco Javier Almendariz-Tapia and Gonzalo Mauricio Figueroa-Torres

*Address all correspondence to: onofrem@iq.uson.mx

Department of Chemical Engineering and Metallurgy, University of Sonora, Hermosillo, Sonora, México

References

[1] Park D, Yun YS, YJM. The past, present, and future trends of biosorption. Biotechnol Bioprocess Eng 2010; 15:86–102.

[2] Watchtower Bible and tract Society of New York, Inc. Peligros del agua contaminada. Despertad Junio 2011; 29.

[3] Monge-Amaya O, Valenzuela J, Acedo-Felix E, Certucha-Barragan M.T, Almendáriz-Tapia F. J. Biosorción de cobre en sistema por lote y continuo con bacterias aerobias inmovilizadas en zeolita natural (clinoptilolita). Rev Int Contam Ambient 2008; 24(3): 107–15.

[4] Cañizares-Villanueva RO. Biosorción de metales pesados mediante el uso de biomasa microbiana. Rev Latinoam Microbiol 2000; 42:131–43.

[5] Cohen Y. Biofiltration – the treatment of fluids by microorganisms immobilized into the filter bedding material: a review. Biores Technol 2001; 77:257–74.

[6] Monge-Amaya O, Valenzuela-García JL, Acedo-Félix E. Biosorción de cobre con bacterias aerobias inmovilizadas en zeolita. Aislamiento, Caracterización, Selección de microorganismos y Cinética de Biosorción. Lambert Academic Publishing. Germany. 2010. pp. 1–136.

[7] Yilmaz EI. Metal tolerance and biosorption capacity of Bacillus circulans strain EB1. Res Microbiol 2003; 154:409–15.

[8] Gómez-Alvarez A, Villalba–Atondo A, Acosta–Ruiz G, Castañeda Olivares M, Kamp, D. Metales pesados en el agua superficial del Río San Pedro durante 1997 y 1999. Rev Int Contam Ambient 2004; 20(1):5–12.

[9] Gómez-Alvarez A. Tesis doctoral. Estudio de caracterización y biodisponibilidad de metales pesados (agua y sedimento) del Río San Pedro, localizado en una región semiárida del noroeste de México. Editorial Universidad Autónoma de Baja California. Instituto de Ingeniería. 2008.

[10] http://www.bbc.com/news/world-latin-america-28850701(19 August 2014 Last updated at 11:02 GMT).

[11] Volke STL, Velasco TJA, De la Rosa PDA, Solórzano OG. Evaluación de tecnologías de remediación para suelos contaminados con metales. INE. Dirección de Investigación en Residuos y Proyectos Regionales. Primer informe. 2004.

[12] Ahalya T, Ramachandra Kanamad R. Biosorption of Heavy Metals. Res J Chem Environ 2003; 7:4.

[13] Reyes E, Cerino E, Suárez HM. Remoción de metales pesados con carbón activado como soporte de biomasa. Ingenierías 2006; IX(31):59–64.

[14] Sag Y, Kutsal T. Recent trends in the biosorption of heavy metals: a review. Biotechnol Bioprocess Eng 2001; 6(6):376–85.

[15] Volesky B. Biosorption of heavy metals. CRC Press, Boca Raton, 1990:1–5.

[16] Chojnacka K. Biosorption and bioaccumulation – the prospects for practical applications. [Review]. Environ Int 2010; 36(3):299–307.

[17] Montalvo S, Guerrero L, Borja R, Sanchez E, Milan Z, Cortes I, de la Rubia MA. Application of natural zeolites in anaerobic digestion processes: a review. Appl Clay Sci 2012; 58:125–33.

[18] Steciow M. (2005). Biorremediación. http://www.cricyt.edu.ar/enciclopedia/terminos/Biorremed.htm (accesed september 2014).

[19] Garbisu C, Amézaga I, Alkorta I. (2002). Biorremediación y Ecología. Ecosistemas http//www.aeet.org/ecosistemas/023/opinion1.htm (accessed June 2010).

[20] Cervantes C. Las relaciones de las bacterias y los metales. Carisma de la Ciencia 2006;(3): 1–4.

[21] Pelczar JM, Reid RD, Chan ECS. Microbiología Mc Graw- Hill México1993;18:119.

[22] Hernandez-Mata K, Monge-Amaya O, Certucha-Barragán MT, Almendariz-Tapia FJ, Acedo-Félix E. Metallic biosorption using yeasts in continuous systems. Int J Photoenergy 2013.1-4.

[23] García-Morales J, Romero-García L, Sales-Márquez D. Influencia de las paradas estacionales sobre la biomasa inmovilizada en reactores anaerobios: Ingeniería del Agua, 1999:6:249–56.

[24] Chiu-Yue L, Chin-Chao C. Effect of heavy metals on the methanogenic UASB granule. Water Res 1999; 33(2):409–16.

[25] Visser, PLEASE ADD FULL REFERENCE

[26] APHA, 1995, Standard Methods: Washington, DC., American Public Health Association.

[27] Volesky B, Weber J, Park JM. Continuous-flow metal biosorption in a regenerable Sargassum column. Water Res 2003; 37(2):297-306.

[28] Figueroa-Torres GM, Certucha-Barragán MT, Almendariz-Tapia FJ, Monge-Maya O, Acedo-Felix E, Pech-Canul MI, et al. Effect of copper and iron on acidogenic biomass in an anaerobic packed bed reactor. Adv Biosci Biotechnol 2014; 5(6):564–571.

[29] Powell GE, Archer DB. Online titration method for monitoring buffer capacity and total volatile fatty acid levels in anaerobic digesters. Biotechnol Bioeng 1989; 33:570–7.

[30] García-Morales J, Romero-García L, Sales-Márquez D. Influencia de las paradas estacionales sobre la biomasa inmovilizada en reactores anaerobios. Ingeniería Del Agua 1999; 6:249–56.

[31] Bhagat M, Burgess JE, Antunes APM, Whiteley CG, Duncan JR Precipitation of mixed metal residues from wastewater utilising biogenic sulphide. Minerals Eng 2004; 17(7–8):925–32.

[32] Kaksonen AH, Plumb JJ, Robertson WJ, Riekkola-Vanhanen M, Franzmann PD, Puhakka JA. The performance, kinetics and microbiology of sulfidogenic fluidized-bed treatment of acidic metal- and sulfate-containing wastewater. Hydrometallurgy 2006; 83(1–4):204–13.

[33] Sampaio RMM, Timmers RA, Xu Y, Keesman KJ, Lens PNL. Selective precipitation of Cu from Zn in a pS controlled continuously stirred tank reactor. J Hazardous Mater 2009; 165(1–3):256–65.

[34] Bijmans MFM. Sulfate reduction under acidic conditions for selective metal recovery. Ph.D. Thesis. Wageningen University, Wageningen, The Netherlands. 2008; 156p.

[35] Liamleam W, Annachhatre AP. Electron donors for biological sulfate reduction. Research review paper. Biotechnol Adv 2007; 25(5):452–63.

[36] Christensen B, Laake M, Lien T. Treatment of acid mine water by sulfate-reducing bacteria: results from a bench scale experiment. Water Res 1996; 30:1617–24.

[37] Diels L, van der Lelie D, Gemoets J, Springael D, Geets J, Vos J, Bastiaens L. In situ bioprecipitation of heavy metals in groundwater. In: Ciminelli VST, Garcia Jr O. (eds.) Biohydrometallurgy: Fundamentals, Technology and Sustainable Development. Process Metallurgy Series; Elsevier Science: New York. 2001; 11B:503–11.

[38] Kaksonen AH, Puhakka J. A sulfate reduction based bioprocesses for the treatment of acid mine drainage and the recovery of metals. Rev Eng Life Sci 2007;7(6):541–64.

[39] Tuppurainen KO, Väisänen AO, Rintala JA. Zinc removal in anaerobic sulphate-reducing liquid substrate process. Minerals Eng 2002; 15(11):847–52.

[40] Alvarez MT, Crespo C, Mattiasson B. Precipitation of Zn(II), Cu(II) and Pb(II) at bench-scale using biogenic hydrogen sulfide from the utilization of volatile fatty acids. Chemosphere 2007; 66(9):1677–83.

[41] Jong T, Parry DL. Adsorption of Pb(II), Cu(II), Cd(II), Zn(II), Ni(II), Fe(II), and As(V) on bacterially produced metal sulfides. J Colloid Interface Sci 2004; 275:61–71.

[42] Yi Z-J, Tan K-X, Tan A-L, Yu Z-X, Wang S-Q. Influence of environmental factors on reductive bioprecipitation of uranium by sulfate reducing bacteria. Int Biodeterior Biodegrad Article in press. 2007; 9p. doi:10.1016/j.ibiod.2007.04.001.

[43] Sirianuntapiboon S, Chairattanawan K, Rarunroeng M. Biological removal of cyanide compounds from electroplating wastewater (EPWW) by sequencing batch reactor (SBR) system. J Hazard Mater 2008; 154(1–3):526–34.

[44] Van Lier JB, Tilche A, Ahring BK, Macarie H, Moletta R, Dohanyos M, Hulshoff-Pol LW. New perspectives in anaerobic digestion. Wat Sci Technol 2001; 43(1):1–18.

[45] Kaksonen AH, Riekkola-Vanhanen ML, Puhakka JA. Optimization of metal sulphide precipitation in fluidized-bed treatment of acidic wastewater. Water Res 2003; 37(2):255–66.

[46] Karri S, Sierra-Alvarez R, Field JA. Toxicity of copper to acetoclastic and hydrogenotrophic activities of methanogens and sulfate reducers in anaerobic sludge. Short Communication. Chemosphere 2006; 62(1):121–7.

[47] Utgikar VP, Chaudhary N, Koeniger A, Tabak HH, Haines JR, Govind R. Toxicity of metals and metal mixtures: analysis of concentration and time dependence for zinc and copper. Water Res 2004; 38:3652–8.

[48] Sani RK, Peyton BN, Brown LT. Copper-induced inhibition of growth of Desulfovibrio desulfuricans G20: assessment of its toxicity and correlation with those of zinc and lead. Appl Environ Microbiol 2001; 67:4765–72.

[49] González-Silva BM. Determinación de parámetros cinéticos y de inhibición por sulfuro en un reactor UASB sulfato-reductor. Tesis de maestría en ciencias aplicadas. División de ciencias ambientales. Instituto Potosino De Investigación Científica Y Tecnológica, A.C. San Luis Potosí, S.L.P. 2007; 61p.

[50] Gupta A, Flora JR, Gupta M, Sayles GD, Suidan MT. Methanogenesis and sulfate reduction in chemostats-I. Kinetic studies and experiments. Water Res 1994; 28:781–93.

[51] Chen BY, Utgikar VP, Harmon SM, Tabak HH, Bishop DF, Govind R. Studies on biosorption of zinc(II) and copper (II) on Desulfovibrio desulfuricans. Int Biodeterior Biodegrad 2000; 46:11–8.

[52] Glombitza F. Treatment of acid lignite mine flooding water by means of microbial sulfate reduction. Waste Manage 2001; 21: 197–203.

[53] Baskaran VK. Kinetics of anaerobic sulphate reduction in immobilised cell bioreactors. Thesis of degree of master of sciences. College of graduate studies and research. Department of chemical engineering. University of Saskatchewan, Saskatoon, SK. Canada. 2005; 148.

[54] Esposito G, Veeken A, Weijma J, Lens PNL. Effect of the use of biogenic sulphide on ZnS precipitation under different process conditions. Sep Purif Technol 2006; 51:31–9.

[55] Lens PNL, Meulepas RJW, Sampaio R, Vallero M, Esposito G. Microbial sulfur metabolism. En: Dahl C, Friedrich CG. (eds.) Bioprocess engineering of sulfate reduction for environmental technology. Chapter 22. Springer-Verlag. Berlin Heidelberg, New York. 2008; 285–295. ISBN-13 978-3-540-72679-1.

[56] Oude-Elferink SJWH, Visser A, Hulshoff-Pol LW, Stams AJM. Sulfate reduction in methanogenic bioreactors. FEMS Microbiol Rev 1994; 15:119–36.

[57] Lettinga G, Van Velsen AFM, Hobma SW, Zeeuw W, Klapwijk A. Use of upflow sludge blanket reactor concept for biological wastewater treatment, especially for anaerobic treatment. Biotechnol Bioeng 1980; 22:699–734.

[58] Lettinga G. Anaerobic digestion and wastewater treatment systems. Antonie van Leeuwenhoek 1995; 67:3–28.

[59] Nielsen J, Vílladsen J. Bioreaction Engineering Principles. Plenum Press. New York and London. 1994.

[60] García-Ochoa E., V. E. Santos J. A. Casas. Chemical structured kinetic model for xanthan production. Enzyme and Microbial Technology 2004. 35, 284–292

[61] Regalado A. Peralta-Reyes E, Gonzalez-Rugerio CA. Como hacer un modelo matemático (Ensayo), Temas de Ciencia y Tecnología 2008; 12(35):9–18.

[62] Laidler MA. Cinética de Reacciones. ED. Alhambra. La Fuensanta, Móstoles (Madrid). 1979; 43–7.

[63] Labuza TP, Riboh D. Theory and application of Arrhenius kinetics to the prediction of nutrients losses in food. Food Technol 1982:66–74.

[64] Wankasi M. Sorption of Pb2+ and Cu2+ ions from aqueous solution by chloride complexation. Soil Sediment Contamin J 2006; 11(6): 841–859.

[65] Volesky B, May-Phillips HA. Biosorption of heavy metals by Saccharomyces cerevisiae. Appl Microbiol Biotechnol 1995; 42(5):797–806.

Chapter 4

Adsorption of Heavy Metal onto the Materials Prepared by Biomass

Naoki Kano

Additional information is available at the end of the chapter

http://dx.doi.org/10.5772/61507

1. Introduction

The amount of heavy metal ions released into the environment has been increased due to industrial activities and technological development. Furthermore, indiscriminate disposal has caused worldwide concern for many years because of the toxicity, accumulation in the food chain, persistence in nature, and concentration by organisms [1–3]. Heavy metals are not biodegradable and tend to accumulate in living organisms, causing various diseases and disorders [4, 5]. It is, therefore, important to reduce the levels of toxic metals or to completely remove them from wastewaters before being discharged into the environment [6]. Then, the minimization and assessment of harmful pollutants such as lead (Pb) and uranium (U) in the environment are very significant from the viewpoint of environmental protection.

There are many processes for the treatment of metal-contaminated wastewaters, including chemical precipitation, membrane filtration, reverse osmosis, ion exchange, and adsorption. However, their use is limited due to various disadvantages [7]. Adsorption has been proved as one of the most efficient methods for the removal of heavy metals from aqueous media [8].

Recently, adsorption based on carbonaceous materials including activated carbon (AC) [9], biochar [10], and carbon nanotubes [11] has been gradually applied to this area. Activated carbon has shown great potential for the removal of various inorganic and organic pollutants and radionuclides due to properties such as large surface area, microporous structure, and high adsorption capacity. As a promising material among nanostructured carbon materials, powdered activated charcoal continues to attract tremendous attention due to its unique physical and chemical properties [3, 12]. In particular, the chemical functionalization of activated carbon can modify its physical and chemical properties, leading to an improved performance in various applications [13–15]. The activation of AC is known to play a key role in enhancing adsorption efficiency via surface morphology modifications with certain functional groups [15–18].

Zeolites are three-dimensional aluminosilicate minerals with a porous structure that have valuable merits, such as cation exchange, molecular sieving, catalysis, and adsorption [19–21]. However, in many cases, these materials do not exhibit high adsorption efficiencies for target metals, and therefore their modification has been reported to enhance their adsorption potential [6, 21–23]. Chitosan, as an abundant natural polysaccharide, has attracted much attention in the biomaterial area because of its biocompatibility, biodegradability, and nonantigenicity [20, 21, 24]. Then, it is expected that the zeolite/chitosan hybrid material, which can incorporate the merits of both materials, may exhibit promising capability for the removal of heavy metal from aqueous solution.

Considering the above, adsorption experiments for heavy metals from aqueous solutions by modified carbon and zeolite/chitosan hybrid material were carried out using Inductively Coupled Plasma Atom Emission Spectrometer (ICP-AES) and ICP-MS in this work to obtain the optimum conditions for heavy metal adsorption process. In this paper, we first present the results of adsorption experiments for U(VI) using "Yukitsubaki" carbon (special product for Aga Town in Niigata Prefecture; abbreviated to YKC herein after) modified with nitric acid as part of the adsorption study of metals using relatively low-cost biomass. Second, the results of adsorption experiments for Pb by activated carbon modified with nitric acid are shown. Then, we finally present the results of adsorption experiments for Pb using zeolite/chitosan hybrid composite (abbreviated to ZCHC herein after).

Our aim was to investigate the efficiency of biomass (or the material prepared by biomass) as an adsorbent for heavy metals for more practical use in the future. Adsorption isotherms of heavy metals were studied and analyzed using Langmuir and Freundlich equations. Furthermore, to evaluate the characteristics of the adsorbents used in this work, the surface morphology, specific surface area, and functional groups of these materials were determined by SEM (Scanning Electron Microscopy), N_2-BET (Brunaeur, Emmet, and Teller) method, and FT-IR (Fourier Transform Infrared Spectroscopy).

2. Experimental Work

2.1. Samples

2.1.1. The charcoal (Yukitsubaki carbon, YKC) modified with nitric acid (HNO_3)

YKC was made from the trunk of *Camellia japonica L.*, which grows in the northeast of Japan, a particularly special product of Aga Town in Niigata Prefecture. YKC was pestled and sieved with 60 mesh. For the removal of ash in YKC, deashing was carried out with boiled ultrapure water. Then, it was dried at 110 °C in an oven. For enhancing the adsorption capacity, YKC carbon was oxidized in 10% HNO_3 and 30% HNO_3 at 90 °C for 4 h and washed with ultrapure water in order to completely remove the residue. Subsequently the sample was heated at 300

°C for 2 h to remove the residual nitrate ions on YKC. In this paper, the pristine YKC, and the modified YKCs oxidized by 10% and 30% HNO_3 were named YKC0, YKC10, and YKC30, respectively.

2.1.2. The Activated Carbon (AC) modified with potassium permanganate ($KMnO_4$)

Activated carbon was washed with deionized water (at 80 °C) to remove fine powders and contaminants, and was then dried at 110 °C for 2 h before use. Five grams of activated carbon was placed in a 200-mL conical flask, which contained 50 mL $KMnO_4$ solution (0.01 or 0.03 mol/L). After adjusting the temperature to 25 °C and stirring for 12 h, the resulting solution was filtrated through a 0.45-μm membrane filter. Then, the filtrate was washed with deionized water until the pH of the filtrate was constant. The activated carbon was dried at 70 °C for 6 h. The untreated and treated (i.e., modified with 0.01 mol/L and 0.03 mol/L $KMnO_4$ solution) activated carbon were described as AC_0, AC_{K1}, and AC_{K3}, respectively.

2.1.3. The Zeolite/Chitosan Hybrid material (ZCHC)

Zeolite was heated at 700 °C for 3 h to activate the surface in a muffle furnace and then washed with hydrochloric acid (5%, volume) and deionized water (at 80 °C) to remove fine powders and contaminants and was then dried at 110 °C for 2 h before use.

ZCHC was prepared by mixing solutions of chitosan and dispersions of zeolite in water. The general procedure of the synthesis is based on sol–gel method [25, 26]. First, 1 g chitosan was dissolved in 20 mL of 0.2 M acetic acid with constant stirring at temperature of 50 °C. Then, 10 mL of deionized water was added into 10 mL chitosan sol solution and was heated and stirred for 1 h. These solutions were mixed, while zeolite was dispersed in the chitosan solution with constant stirring for 5 h at a temperature of 50 °C. Then, the solution was transferred into five 10-mL centrifuge tubes, which were centrifuged at 9000 rpm for 5 min, and then washed with deionized water to remove contaminants. The mixed solutions were put on Petri dishes and were left to dry at room temperature for 24 h. The obtained films with a thickness of 0.1 mm were used for the following adsorption experiments.

2.2. Adsorption experiment for heavy metals using natural materials prepared by biomass

The following adsorption experiments were performed in a batch system using the above-mentioned samples. Experimental conditions (i.e., pH, contact time, sorbent amount, and temperature) in this work were optimized and determined based on our preliminary experiments and other studies [15, 20, 22]. The pH of each solution was adjusted by using 0.1 molL $^{-1}$ NH_3 aq/0.1 molL^{-1} HNO_3. Adsorption isotherms of heavy metals onto these materials were measured at varying initial concentrations under optimized condition.

2.2.1. The charcoal (Yukitsubaki carbon, YKC) modified with nitric acid (HNO_3)

Each YKC sample of 30 mg was contacted with 50 mL of a known amount of U(VI) in a 100-mL conical flask, and the suspensions were shaken in a water bath. Adsorption experiments were conducted in the pH range of 2–10, with an adsorbent dosage of 0.1–0.6 gL^{-1}, at a temperature of 5–45 °C, contact time from 1 to 24 h, and an initial concentration from 0 to 200 μgL^{-1}.

2.2.2. The Activated Carbon (AC) modified with potassium permanganate ($KMnO_4$)

Activated carbon was thoroughly mixed with 50 mL of a known amount of Pb^{2+} in a 200-mL conical flask, and the suspensions were shaken in a water bath at room temperature (25 ± 2 °C). Adsorption experiments were conducted in the pH range of 3–7, with an adsorbent dosage of 0.1–1.5 gL^{-1}, contact time from 1 to 24 h, and an initial concentration from 20 to 200 mgL^{-1}.

2.2.3. The zeolite/chitosan hybrid material (ZCHC)

Each ZCHC was thoroughly mixed with 50 mL of a known amount of Pb^{2+} in a 200-mL conical flask, and the suspensions were shaken in a water bath at room temperature (25 ± 2 °C). Adsorption experiments were conducted in the pH range of 3–7, a contact time from 1 to 48 h, and an initial concentration from 20 to 200 mgL^{-1}, with an adsorbent dosage of 1.0 gL^{-1}.

Following each sorption experiment, the suspension containing modified carbon (or ZCHC) and heavy metal solution was filtered through a 0.45-μm membrane filter (Advantec Mixed Cellulose Ester, 47 mm) to remove heavy metals that have been adsorbed into the adsorbent, and the concentration of these metals in the filtrate was determined with ICP-MS (X2, ThermoFisher) or AAS (Z-5000, HITACHI).

The metal uptake by each adsorbent was calculated using the following equation:

$$q = \frac{(C_i - C_e)}{m} \cdot V \tag{1}$$

where q is the adsorption capacity of heavy metal with the adsorbent at equilibrium (mg g^{-1}), C_i and C_e are the initial and equilibrium concentrations of heavy metal in a batch system, respectively (mg L^{-1}), V is the volume of the solution (L), and m is the weight of the adsorbent (g).

2.3. Langmuir and Freundlich isotherm model

Adsorption isotherms are commonly used to reflect the performance of adsorbents in adsorption processes. To examine the relationship between the metal uptake (q_e) and the concentration of metal ions (C_e) at equilibrium, adsorption isotherm models are widely used for fitting data.

The Langmuir model assumes monolayer adsorption on a surface given by the following equation:

$$\frac{C_e}{q_e} = \frac{C_e}{q_{max}} + \frac{1}{K_L q_{max}} \tag{2}$$

where C_e is the concentration of U(VI) (or Pb^{2+}) in a batch system at equilibrium (mg L^{-1}), q_e is the amount of U(VI) (or Pb^{2+}) adsorption at equilibrium (mg g^{-1}), q_{max} is the maximum adsorption capacity on the surface of activated carbon (or ZCHC) (mg g^{-1}), and K_L is the Langmuir adsorption constant (L mg^{-1}) [27, 28]. A plot of C_e/q_e versus C_e gives a straight line with a slope of $1/q_{max}$, and an intercept of $1/(K_L q_{max})$.

The Freundlich equation is widely used in the field of environmental engineering and was applied based on the work by Dahiya et al. [1, 29]. Freundlich isotherm can also be used to explain adsorption phenomenon as given below:

$$\log_{10} q_e = \log_{10} K_F + (1/n)\log_{10} C_f \tag{3}$$

where K_F and n are constants incorporating all factors affecting the adsorption capacity and an indication of the favorability of metal ion adsorption onto a biosorbent, respectively. It is shown that $1/n$ values between 0.1 and 1.0 correspond to beneficial adsorption. That is, q_e versus C_f in a log scale can be plotted to determine the values of $1/n$ and K_F.

2.4. Kinetic model

Kinetics of heavy metal uptake was modeled using the pseudo-first-order and pseudo-second-order Lagergren equations [30, 31]. The pseudo-first-order reaction of Lagergren for sorption can be expressed as follows based on Bhat et al. [32]:

$$\mathrm{d}Q / \mathrm{d}t = k_1(Qe - Qt) \tag{4}$$

where Qe and Qt are the amount of metal adsorbed per unit weight (μg/g) of sorbent at equilibrium and at any time t (min), respectively, and k_1 is the rate constant of pseudo-first-order sorption (min^{-1}). The integrated form of the above equation after applying the boundary conditions, for $t = 0$, $qt = 0$, becomes

$$\log(Qe - Qt) = \log(Qe) - (k_1 / 2.303)t \tag{5}$$

The value of the rate constant (k_1) and Qe for the pseudo-first-order sorption reaction can be obtained by plotting log(Qe −Qt) versus t.

The pseudo-second-order rate of Lagergren can be expressed as follows:

$$\mathrm{D}Q / \mathrm{d}t = k_2(Qe - Qt)^2 \tag{6}$$

where k_2 (g/(μg min)) is the rate constant for the pseudo-second-order sorption. The integrated linear form of Eq. (9) can be represented as follows:

$$t/Q = 1/k_2Q^2_e + (1/Q_e)t \quad (7)$$

The pseudo-second-order rate constant (k_2) and Q_e can be calculated from the intercept and slope of the linear plot of t/Qt versus t.

The initial adsorption rate (h) can be determined from k_2 and Qe values using

$$h = k_2Qe^2 \quad (8)$$

3. Results and discussion

3.1. The charcoal (Yukitsubaki carbon, YKC) modified with nitric acid (HNO_3)

3.1.1. Characterization of YKC

The morphologies of pristine and modified YKCs characterized by SEM (JCM-6000, JEOL) are shown in Figure 1. Moreover, the surface properties including specific surface areas of pristine and modified YKCs determined by N_2-BET method (TriStar II 3020, Micromeritics) are shown in Table 1. The pore structures of all the samples are similar to each other. However, judging from the SEM image shown in Figure 1, the surface area of YKC30 (Figure 1C) seems to be slightly changed with the acid treatment, whereas that of YKC10 (Figure 1B) seems to be hardly varied. This is consistent with the data of the specific surface area shown in Table 1. The decrease in the specific surface area of YKC30 would be attributable to the excessive oxidation with a high concentration of nitric acid.

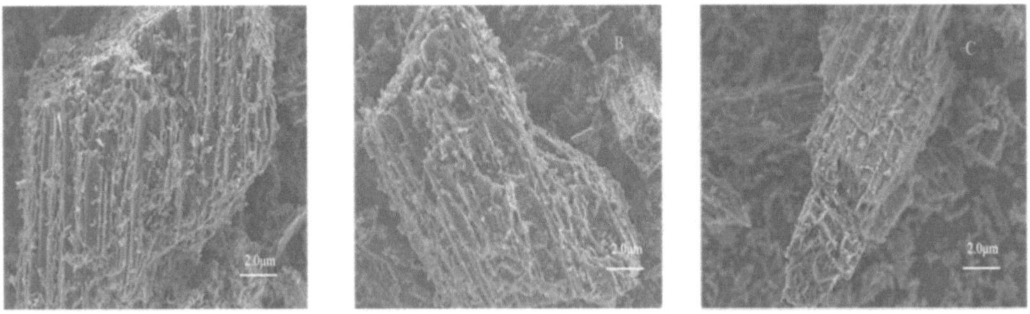

A B C

Figure 1. The SEM images of pristine and modified YKCs. (A: YKC0, B: YKC10, C: YKC30)

The FT-IR (FTIR-4200, Jasco) spectra of pristine and modified YKCs are shown in Figure 2. From the figure, a characteristic broadband, which may be due to graphite structure in YKCs,

Samples	Specific surface area (m^2/g)	Pore volume (cm^3/g)	Pore size (nm)
YKC0	157	0.0854	2.18
YKC10	158	0.0852	2.16
YKC30	144	0.0774	2.15

Table 1. Surface properties of pristine and modified YKCs

is observed at around 1610 cm^{-1}. The peak at 3300 cm^{-1} is related to hydroxyl groups (–OH), and 1120 cm^{-1} is related to carbonyl groups (–C=O), and then the peaks at 2920 and 2850 cm^{-1} are associated with C–H [33–36]. The results of FT-IR analysis show that some kinds of functional groups (such as carbonyl groups and hydroxyl groups) are introduced to YKC surfaces successfully by oxidation.

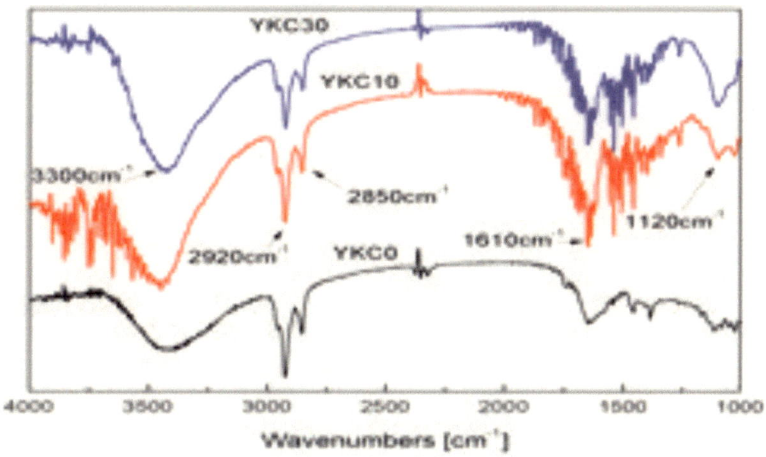

Figure 2. The FT-IR spectra of pristine and modified YKCs

3.1.2. Influence of parameters on adsorption

3.1.2.1. Contact time

The effect of contact time on U(VI) adsorption was investigated to study the adsorption rate of U(VI) removal. The percentage removal of U(VI) for the concentration of 100 μg/L with the adsorbent dosage of 30 mg at a pH of 3.0 is shown in Figure 3.

The adsorption equilibrium of U(VI) on YKC10 was reached within 30 min, much faster than that on YKC0 and YKC30. For the rest of the study, 480 min (8 h) was selected as the contact time.

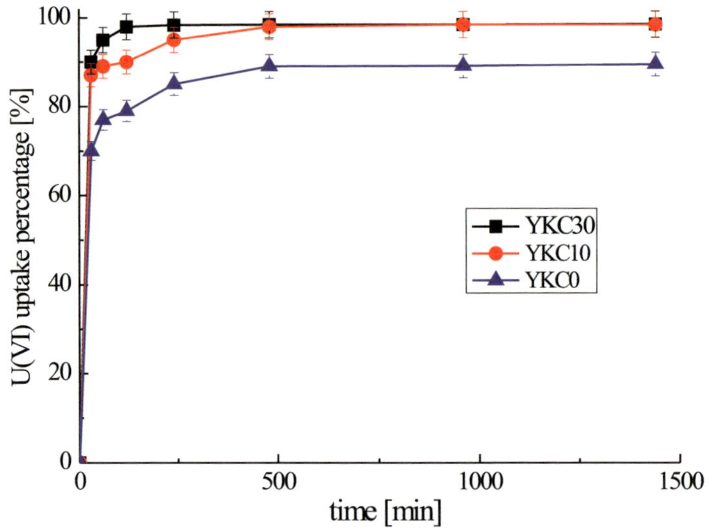

Figure 3. Effect of contact time on U(VI) adsorption (pH = 3.0, *T* = 25 °C, and adsorbent dose of 30 mg)

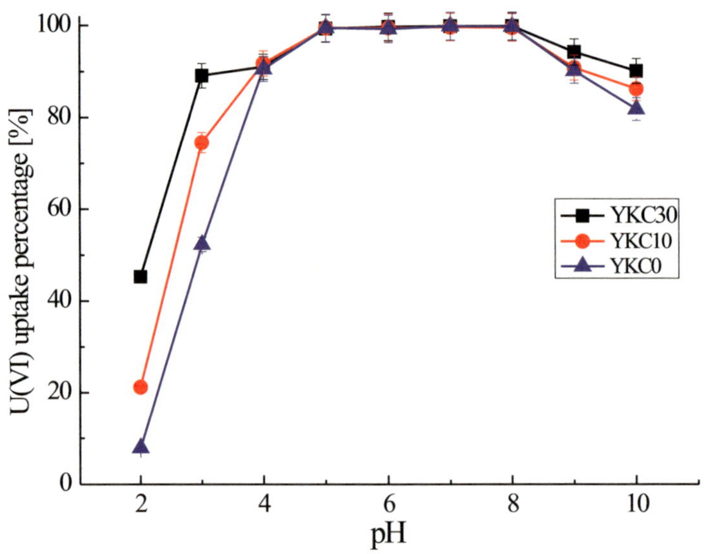

Figure 4. Effect of pH on U(VI) adsorption (contact time of 8 h, *T* = 25 °C, and adsorbent dose of 30 mg)

The adsorption of U(VI) on the YKC as a function of pH is shown in Figure 4. From this figure, it can be seen that the amount of U(VI) uptake increased with increasing solution pH. The highest uptake was observed at a pH of 5–8 for all YKCs, and the uptake of U(VI) decreased slightly with increasing pH at a pH of >8. The difference in the pH dependence is not clearly found among YKCs, although the difference of adsorption capacities among YKCs was obviously found. It is known that U exists in different forms depending on the pH. U exists predominantly as monomeric species UO_2^{2+}, and small amounts of $UO_2(OH)^+$ at a pH of ≤4.3, and at a pH of ≥5, colloidal or oligomeric species, i.e., $(UO_2)_2(OH)_2^{2+}$, $(UO_2)_3(OH)_5^+$, $(UO_2)_4(OH)_7^+$, and $(UO_2)_3(OH)_7^-$, are formed [37–40].

That is, U usually exists as cationic species in solution at a pH of 5. Furthermore, from the results of the FT-IR spectra of YKC shown in Figure 2, hydroxyl groups (–OH) are introduced onto YKC.

From the above, it can be considered that the U(VI) adsorption occurred dominantly by the cation exchange reaction between the H^+ of hydroxyl groups on YKC and the cationic species of U(VI).

Then, a pH of 5 was assumed for further experimental work, although the uptake of U(VI) maintains the highest level at a pH of 5–8.

3.1.2.2. *Adsorbent amount and adsorption temperature*

Under optimized conditions of pH (i.e., pH 5) and contact time (i.e., 8 h), the adsorption behavior at different adsorbent amounts (10–60 mg) or adsorption temperatures (5–45°C) has been studied in 50 mL of a 100-μg/L aqueous U(VI) solution in 100-mL flasks.

From the above experiment, the adsorbent amount was selected as 30 mg in our work in consideration of the economic cost, although the adsorption efficiency slightly increased with an increase in the adsorbent amount. On the other hand, 25 °C was selected because the highest uptake was observed at 25 °C, and there is no appreciable change in the uptake of U(VI) with increasing temperatures.

3.1.3. *Adsorption isotherm study*

The Langmuir and Freundlich isotherms for the adsorption of U(VI) onto YKC are given in Figures 5 and 6.

The linear correlation coefficients (R^2) of Langmuir and Freundlich isotherms for U(VI) using pristine and modified YKCs are shown in Table 2 along with other parameters.

From this table, it is found that the R^2 value for each datum is comparatively large for both Langmuir and Freundlich isotherms. Furthermore, it is noted that R^2 values for these data are larger for the Langmuir isotherm than for the Freundlich isotherm. This result suggests that the adsorption on these samples mainly occurred by a monolayer reaction.

From the data of Q_m in Table 2, it is also found that the adsorption capacity of U(VI) on modified YKC is much higher than that on pristine YKC. The results indicated that acid treatment is effective for enhancing the adsorption capacity of U(VI).

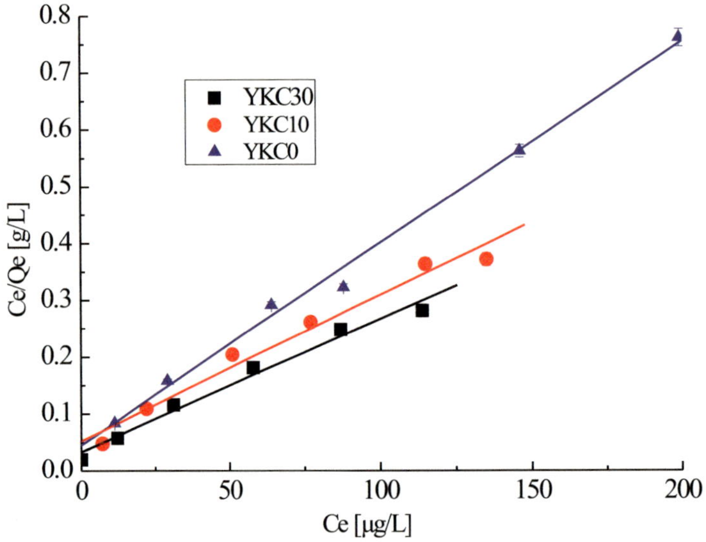

Figure 5. The Langmuir isotherm of U(VI) adsorption (pH = 5.0, T = 25 °C, and adsorbent dose of 30 mg)

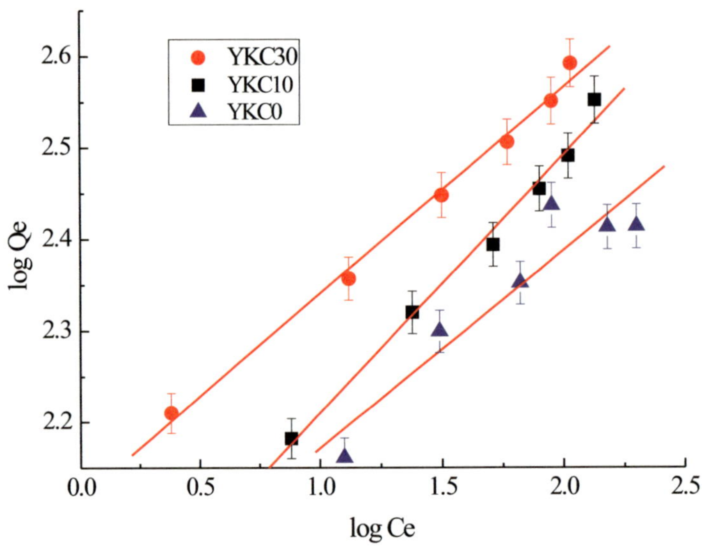

Figure 6. The Freundlich isotherm of U(VI) adsorption (pH = 5.0, T = 25 °C, and adsorbent dose of 30 mg)

	Langmuir isotherm			Freundlich isotherm		
	Q_m(μg/g)	K_L (L/μg)	R^2	b_F	K_F (μg/g)	R^2
YKC0	270	0.112	0.996	0.196	99.4	0.912
YKC10	417	0.0931	0.984	0.222	132	0.891
YKC30	385	0.0522	0.976	0.274	88.3	0.843

Table 2. Coefficients of Langmuir and Freundlich isotherm models

Here, Q_m is the maximum adsorption capacity of the U(VI) (μg/g), K_L is the Langmuir binding constant, which is related to the energy of adsorption (L/μg), and K_F and b_F are the Freundlich constants related to the adsorption capacity and intensity.

3.2. The activated carbon modified with potassium permanganate ($KMnO_4$)

3.2.1. Characterization of the modified activated carbon

The FT-IR spectra of the pristine and modified activated carbon (i.e., AC_0, AC_{K1}, and AC_{K3}) are shown in Figure 7. The pristine and modified activated carbon displayed the characteristic bands of the graphite structure of carbon at 1615 cm^{-1} [8, 15, 41]. Moreover, an OH stretching band, one of the typical peaks of activated carbon, was found at 3300–3500 cm^{-1}. The peak at 3433 cm^{-1} was related to the hydroxyl groups (–OH) stretch from deprotonated pristine and modified activated carbon. The wide peak at 1550–1750 cm^{-1} shows the asymmetric stretch of the carboxylate (–COO–) group [15].

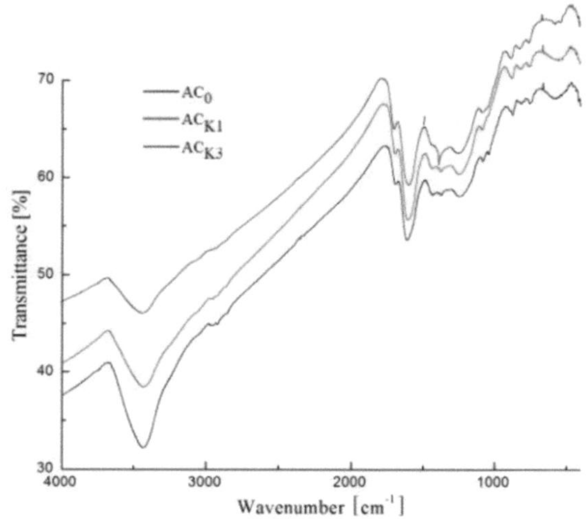

Figure 7. FT-IR spectra of AC_0, AC_{K1}, and AC_{K3}.

The surface properties of the activated carbon were investigated by N_2 adsorption (TriStar II 3020 Micromeritics), and the analytical results for the adsorption/desorption isotherms are shown in Table 3.

Adsorbent	BET surface area ($m^2 \cdot g^{-1}$)	Pore volume ($cm^3 \cdot g^{-1}$)	Pore size (nm)
AC_0	381	0.402	4.23
AC_{K1}	373	0.390	4.18
AC_{K3}	346	0.348	4.03

Table 3. Textural characteristics of activated carbon

The pore volume was calculated from the amount of N_2 adsorbed at the relative pressure of 0.99. The pore size was calculated from the adsorption average pore width (4V/A by BET) in this work. From Table 3, it is found that the pore volume and pore size as well as the specific surface area decreased significantly after modification with $KMnO_4$. The isotherm showed a type H1 isotherm with a clear hysteretic loop, characteristic of disordered microporous materials.

The SEM micrographs of the activated carbon are shown in Figure 8. The modified AC (Figures 8B and 8C) seemed to exhibit a more compact stacking morphology than the pristine AC (Figure 8A), due to cohesive forces, which may be generated from the introduction of oxygen-containing functional groups. These results are consistent with those of the N_2 adsorption–desorption experiment. The decrease of the pore volume and pore size may be related to the increase of acidic groups on the surface of activated carbon treated with $KMnO_4$.

Figure 8. SEM micrographs of the surface of activated carbon: (A) unmodified; (B) modified with 0.01 mol/L $KMnO_4$; (C) modified with 0.03 mol/L $KMnO_4$

3.2.2. Adsorption of lead on modified activated carbon

3.2.2.1. Effect of pH

To investigate the effect of solution pH on Pb^{2+} adsorption efficiency, the pH of the solution was varied from 3 to 7, while the Pb^{2+} concentration was kept constant at 100 mg L^{-1}. The experimental results are presented in Figure 9. The Pb^{2+} adsorption efficiency was at pH 5 regardless of the kind of adsorbent (Figure 9). The uptake of Pb^{2+} increased from 50.8% at pH 3 to 90.0% at pH 5, and at higher pH values, it remained almost constant (or decreased only slightly). Notably, the adsorption capacities decreased at low pH values due to the competition of protons with metal ions for active binding. On the other hand, lead precipitated from the solution at higher pH values as lead hydroxide [1]. From the FT-IR spectra of AC (Figure 7), it was clear that the hydroxyl groups (–OH) were introduced onto AC. We hypothesized that the Pb^{2+} adsorption occurred predominantly by cation exchange reaction between the H^+ of the hydroxyl groups on modified AC and cationic Pb^{2+} species. However, it is possible that Pb^{2+} was removed to some extent via precipitation at higher pH values rather than by adsorption on the modified AC. Hence, pH 5 was used for further experiments.

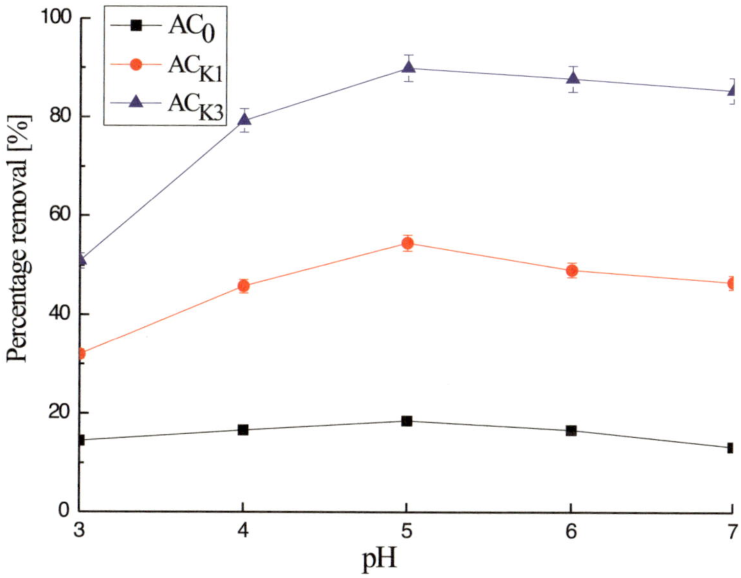

Figure 9. Effect of pH on the removal of Pb^{2+} (%) using modified activated carbon.

3.2.2.2. Effect of contact time

The effect of contact time on Pb^{2+} adsorption efficiency using 1.0 gL^{-1} AC_{K3} (100 mgL^{-1} of solution) was investigated at pH 5.

More than 80% Pb was removed within 1 h, and it gradually increased at 2 h. Approximately 90% of Pb was removed from the solution at the contact time of 2 h. After 2 h, there was no appreciable change. Therefore, 2 h was chosen as the optimized contact time for the rest of the experimental work.

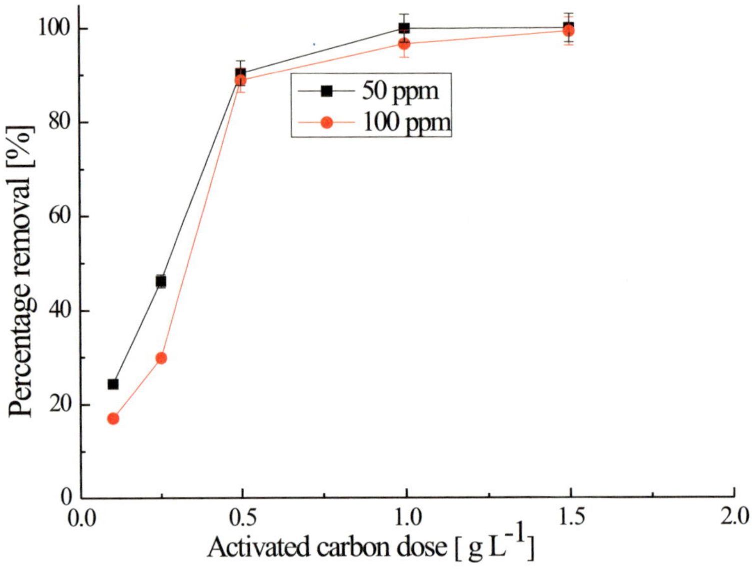

Figure 10. Effect of sorbent dosage on percent removal of Pb^{2+} using modified activated carbon

3.2.2.3. Effect of sorbent dosage

Under the optimized pH conditions (i.e., pH 5) and contact time (i.e., 2 h), the adsorption behavior of AC_{K3} at different dosages (from 0.1 to 1.5 gL^{-1}) was studied in 100 mgL^{-1} Pb^{2+} solution.

More than 90% of Pb^{2+} was removed with a dosage of 1.0 gL^{-1} (Figure 10). The removal increased remarkably with higher dosage rates, but no remarkable increase was observed at dosages greater than 1.0 gL^{-1}. Therefore, 1.0 gL^{-1} was considered as the optimum dosage for the remainder of the study.

3.2.2.4. Effect of competitive ions

Competitive experiments were conducted under the optimized pH conditions (i.e., pH 5), contact time (i.e., time 2 h), and sorbent dosage (i.e., 1 g/L) using different concentrations of Na^+, K^+, Ca^{2+}, or Mg^{2+} separately and combination of all four ions (i.e., 0, 10, 20, 50, 100, 200, and 500 mgL^{-1}). The percent removal of Pb^{2+} decreased in the presence of Na^+, K^+, Ca^{2+}, or Mg^{2+} with concentrations from 0 to 500 mgL^{-1} (Figure 11). A remarkable decrease in the adsorption capacity of Pb^{2+} was not observed, even with common ions at concentrations of 100 mgL^{-1} (i.e., more than 80% Pb^{2+} was removed; Figure 5). This implied that the activated carbon was an efficient adsorbent for Pb^{2+}, although further investigations are required for the realization of practical application.

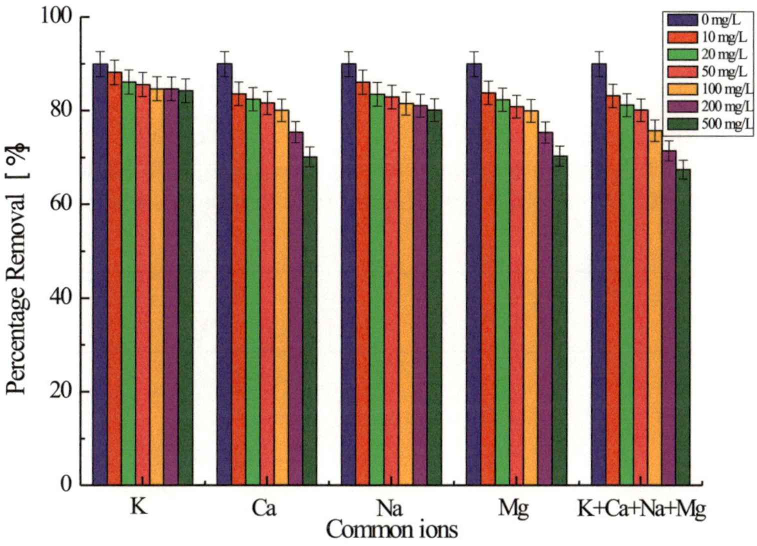

Figure 11. Effect of competitive ions on percent removal of Pb^{2+} using modified activated carbon.

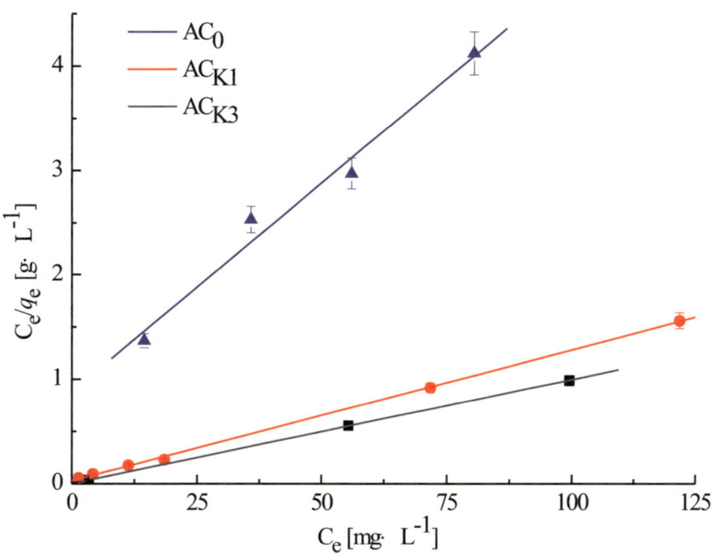

Figure 12. Langmuir isotherm of Pb^{2+} adsorption on AC.

3.2.3. *Adsorption isotherms*

Adsorption isotherms are commonly used to reflect the performance of adsorbents in adsorption processes. The Langmuir adsorption isotherm was applied to the data obtained in this work. A plot of C_e/q_e versus C_e based on the Langmuir model is shown in Figure 12.

From these data, the q_{max} and K_L of AC_0, AC_{K1}, and AC_{K3} were calculated and are shown in Table 4 along with the R^2 (correlation coefficient). The adsorption capacity of AC modified with 0.03 mol/L $KMnO_4$ was about 4 times higher than that of pristine AC.

Adsorbent	q_{max} *(mg·g⁻¹)*	***KL*** **(L·mg⁻¹)**	***R***
AC_0	25.1	0.0452	0.983
AC_{K1}	80.0	0.435	0.998
AC_{K3}	101	4.60	1.00

q_{max}: the maximum adsorption capacity on the surface of activated carbon (mg g^{-1});

K_L: the Langmuir adsorption constant (L mg^{-1});

R: the correlation coefficient.

Table 4. Coefficient of Langmuir isotherm for Pb^{2+} using AC

The R^2 value for each adsorbent was comparatively large (Table 4). Furthermore, the R^2 value for modified AC (i.e., AC_{K1} and AC_{K3}) was particularly large. That is to say, a favorable adsorption of Pb^{2+} by activated carbon was apparent. This result suggested that the adsorption of these samples primarily occurred via a monolayer reaction.

Biosorbent	**Maximum adsorption capacity (mg·g⁻¹)**	**Reference**
Acidified MWCNTs	49.7	[42]
Cicer arientinum biomass	27.8	[7]
Sawdust of Meranti wood	37.2	[43]
Activated carbon prepared from apricot stone	22.9	[16]
Peanut husks carbon	70.0	[44]
Activated carbon modified with KMnO4	101	Present study

Table 5. Comparison of Pb^{2+}adsorption capacities of modified AC with other adsorbents

Next, the adsorption capacity of Pb^{2+} by modified AC was compared with other adsorbents under ambient conditions previously reported in the literature. As shown in Table 5, the maximum adsorption capacity of modified AC toward Pb^{2+} was higher than that of other adsorbents.

The adsorption capacity of activated carbon may be dependent on its surface properties ([7], [45]). Therefore, the modification with $KMnO_4$ (an efficient oxidizer) could change the surface structure of activated carbon to enhance the adsorption capacity of AC toward metals.

3.2.4. Kinetic studies

The linear plot of t/q_t versus t for metal adsorption under the optimized experimental conditions is shown in Figure 13. The coefficient of determination was more than 0.996 of Pb^{2+} on modified AC. The pseudo-second-order rate constant (k) and the amount of adsorbed lead (q_e) obtained from the intercept and slope of the plot of t/q_t versus t are listed in Table 6 along with the regression coefficient (R^2).

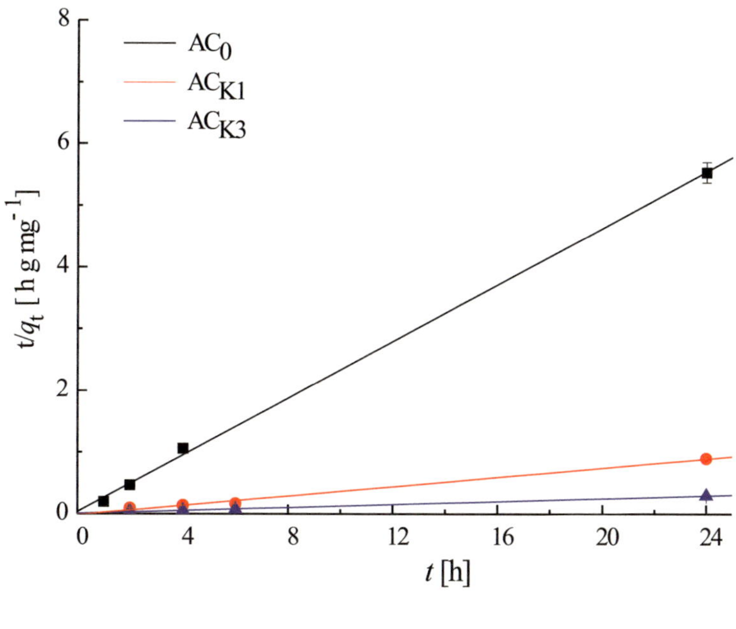

Figure 13. The pseudo-second-order kinetic model for AC.

The adsorption kinetics based on the experimental values were in good agreement with the pseudo-second-order kinetic model. The intraparticle diffusion model indicated that the relationship between the concentration of Pb^{2+} and the square root of time was linear. This suggested that the adsorption process could be controlled by intraparticle diffusion. Moreover, the adsorption at higher temperatures became more dependent on intraparticle diffusion, the rate-determining step.

Activated carbon	qe (mg·g^{-1})	k (g·mol^{-1}·h^{-1})	R
AC_0	26.4	2.87×10^{-3}	0.999
AC_{K1}	81.8	1.16×10^{-2}	0.999
AC_{K3}	101	4.51×10^{-2}	0.999

q_e: the amount of adsorbed lead on the surface of activated carbon at equilibrium (mg g^{-1});

k : the rate constant of the pseudo-second-order adsorption (g mol^{-1} h^{-1});

R: the correlation coefficient.

Table 6. Kinetic coefficients for Pb^{2+} adsorption on AC

3.3. The zeolite/chitosan hybrid material (ZCHC)

3.3.1. Characterization of ZCHC

The surface properties of ZCHC before and after Pb^{2+} adsorption were investigated by N_2 adsorption (TriStar II 3020, Micromeritics), and the analytical results for the adsorption/desorption isotherms are shown in Table 7.

Adsorbent	BET surface area (m^2·g^{-1})	Pore volume (cm^3·g^{-1})	Pore size (nm)
ZCHC (before adsorption)	9.25	0.0485	21.0
ZCHC (after adsorption)	9.19	0.0479	20.7

Table 7. Textural characteristics of ZCHC

The pore volume was calculated from the amount of N_2 adsorbed at the relative pressure of 0.99. The pore size was calculated from the adsorption average pore width (4V/A by BET) in this work.

The SEM micrographs of the ZCHC before and after Pb^{2+} adsorption ZCHC are shown in Figure 14. From Figure 14, it is found that the morphology of ZCHC surface has hardly changed even after exposing Pb^{2+}, although the SEM picture after Pb^{2+} adsorption slightly exhibits a more compact stacking morphology than that before adsorption. From the above observation, ZCHC should be predicted to withstand the repeated use, and hence it can be a good adsorbent for heavy metals such as Pb^{2+}. These results are consistent with those of the N_2 adsorption–desorption experiment.

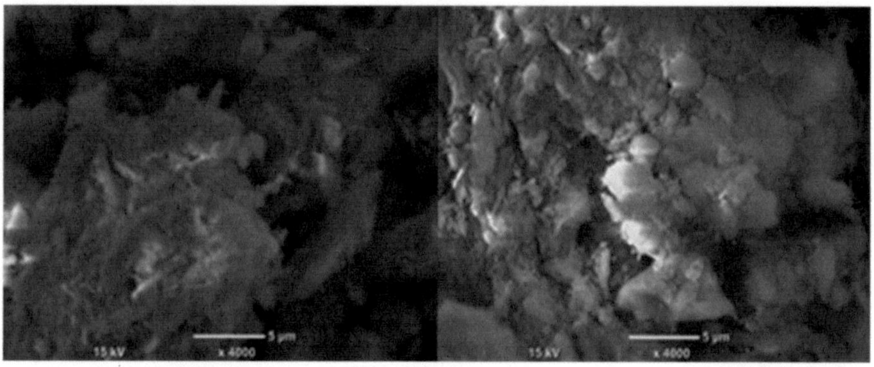

(A) (B)

Figure 14. SEM micrographs of the surface of ZCHC: (A) before Pb^{2+} adsorption; (B) after Pb^{2+} adsorption.

3.3.2. Adsorption of lead on zeolite and ZCHC

With the aim of obtaining the optimum conditions, the effects of pH value and contact time in the case of a fixed dosage of adsorbent (i.e., 1.0 g L^{-1}) on the removal of Pb^{2+} from the aqueous solution were investigated. The effect of pH on Pb^{2+} adsorption onto zeolite and ZCHC is shown in Figure 15. From Figure 15, the uptake of Pb^{2+} increased from 56.7% at pH 3 to 95.6% at pH 5, and at higher pH value, it remained almost constant (or decreased slightly). Notably, the adsorption capacities decreased at low pH values due to the competition of protons with metal ions for active binding. On the other hand, lead precipitated from the solution at higher pH values as lead hydroxide. Hence, pH 5 was used for further experiments.

The effect of contact time on Pb^{2+} adsorption onto zeolite and ZCHC is shown in Figure 16. Approximately 90% of Pb was removed from the solution with ZCHC at the contact time of 1 h, and it gradually increased at 4 h as shown in Figure 16. More than 95% of Pb was removed from the solution at the contact time of 4 h. After 4 h, there was no appreciable change. On the other hand, a contact time of 8 h was needed to attain at equilibrium in case of zeolite. Therefore, 8 h was chosen as the optimized contact time for more certainty for these samples at the rest of the experimental work.

3.3.3. Adsorption isotherms

Adsorption isotherms are commonly used to reflect the performance of adsorbents in adsorption processes. The Langmuir adsorption isotherm was applied to the data obtained in this work. A plot of C_e/q_e versus C_e based on the Langmuir model is shown in Figure 17.

As shown in Table 8, *R* value for each adsorbent was comparatively large, and a favorable adsorption of Pb^{2+} by these samples was apparent. This result suggested that the adsorption of these samples primarily occurred via a monolayer reaction. Furthermore, the maximum adsorption capacity of ZCHC toward Pb^{2+} was higher than that of zeolite.

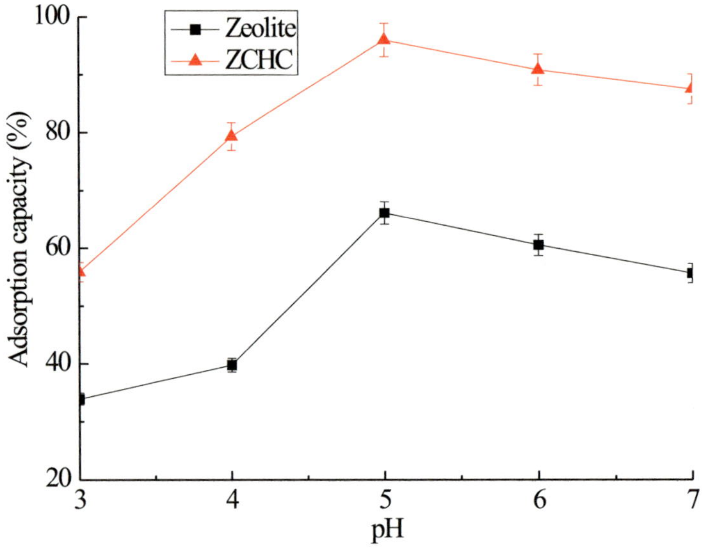

Figure 15. Effect of pH on the removal of Pb^{2+} removal of using zeolite and ZCHC. Pb^{2+} using zeolite and ZCHC.

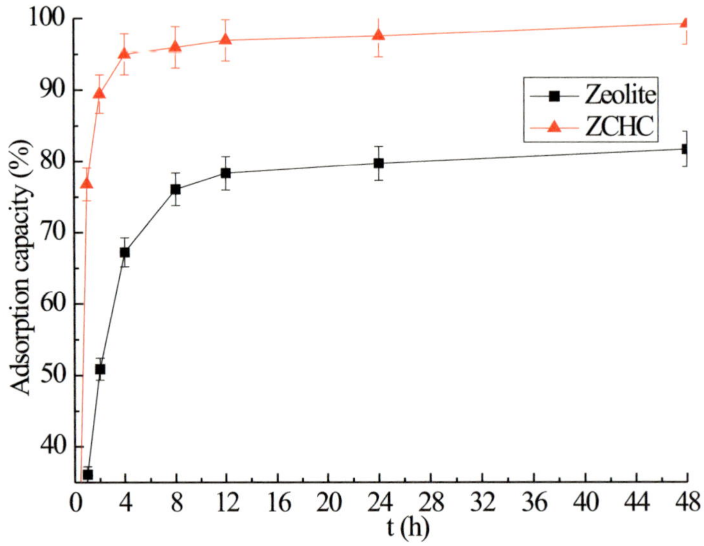

Figure 16. Effect of contact time on the

http://dx.doi.org/10.5772/61507

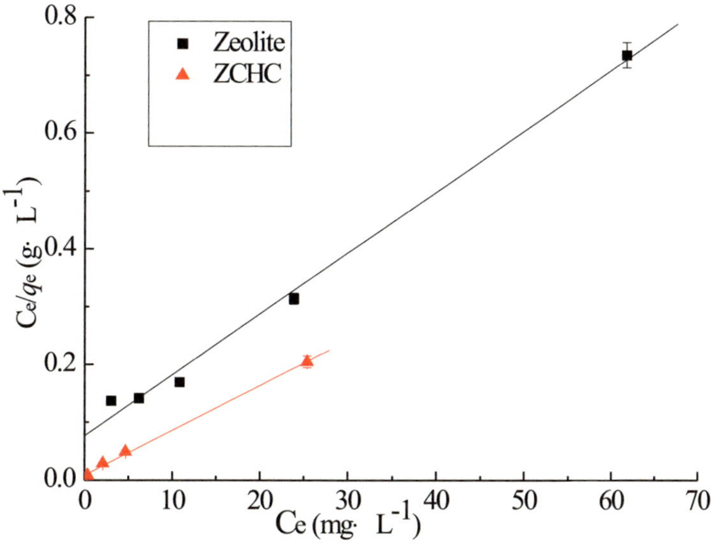

Figure 17. Langmuir isotherm of Pb^{2+} adsorption on zeolite and ZCHC.

Adsorbent	q_{max} $(mg{\cdot}g^{-1})$	*KL* ($L{\cdot}mg^{-1}$)	*R*
Zeolite	66.5	0.196	0.997
ZCHC	139	0.844	0.999

q_{max}: the maximum adsorption capacity (mg g^{-1});

K_L: the Langmuir adsorption constant (L mg^{-1});

R: the correlation coefficient.

Table 8. Coefficient of Langmuir isotherm for Pb^{2+} using zeolite and ZCHC

4. Conclusion

Yukitsubaki carbon (charcoal) modified with nitric acid (10% and 30%), activated carbon modified with potassium permanganate ($KMnO_4$; 0.01 mol/L and 0.03 mol/L), and zeolite/chitosan hybrid composite (ZCHC) prepared with sol–gel method were used as adsorbents for heavy metal ions in this work. Modified carbon treated with HNO_3 or $KMnO_4$ exhibits high ability of chemical adsorption and stronger chemical affinity than pristine carbon. ZCHC prepared in this work also could be suitable as sorbent materials for the removal of heavy metals from aqueous solutions. The adsorbent showed excellent adsorption capacity for Pb^{2+} under our experimental condition, even in the presence of diverse ions (Ca^{2+}, Mg^{2+}, Na^{+}, and

K^+) up to the concentration of 100 $mgdm^{-3}$. It is indicated that these materials used in this work could be effective adsorbents for practical use in the future.

It is also suggested that the removal of U and Pb from aqueous solutions by these materials is mainly due to monolayer sorption because of well fitting for Langmuir model. Furthermore, the best fit was obtained with a pseudo-second-order kinetic model while investigating the adsorption kinetic of carbon modified with HNO_3 or $KMnO_4$.

The data obtained and the method used in this work can be a useful tool from the viewpoint of environmental protection and resource recovery in future work.

Acknowledgements

The present work was partially supported by a Grant-in-Aid for Scientific Research (Research Program (C), No. 25340083) of the Japan Society for the Promotion of Science. This research was also supported by a fund for the promotion of Niigata University KAAB Projects from the Ministry of Education, Culture, Sports, Science and Technology, Japan.

The present work was partially supported by a Grant-in-Aid for Scientific Research of the Japan Society for the Promotion of Science. Yukitsubaki carbon samples were supplied by Mr. J. Sakai of the Facility of Engineering in Niigata University. The authors are also grateful to Mr. N. Saito, Mr. T. Nomoto, Prof. T. Tanaka, Mr. T. Hatamachi, Dr. M. Teraguchi and Prof. T. Aoki of the Facility of Engineering in Niigata University for permitting the use of SEM, Surface Area Analyzer, and FT-IR, and thanks to Dr. K. Fujii and Mr. M. Ohizumi of the Office for Environment and Safety in Niigata University for permitting the use of ICP-MS and for providing helpful advice in measurement.

Author details

Naoki Kano*

Address all correspondence to: kano@eng.niigata-u.ac.jp

Department of Chemistry and Chemical Engineering, Faculty of Engineering, Niigata University, Nishi-Ku, Niigata, Japan

References

[1] Dahiya S., Tripathi R. M., Hegde A. G. (2008) Biosorption of Heavy Metals and Radionuclide from Aqueous Solutions by Pre-treated Arca Shell Biomass. J. Hazard. Mater., 150: 376–386.

[2] Sachdeva S., Kumar A. (2009) Preparation of nanoporous composite carbon membrane for separation of rhodamine B dye. J. Membr. Sci., 329: 2–10.

[3] Wang J., Chen C. (2009) Biosorbents for heavy metals removal and their future. Biotechnol. Adv., 27: 195–226.

[4] Shashikant R.M., Rajamanya V. S. (2003) Biosorption studies of chromium and pentachlorophenol from tannery effluents. Bioresour. Technol., 98: 1128–1132.

[5] Pillai S. S., Mullassery D. M., Fernandez B. N., Girija N., Geetha P., Koshy M. (2013) Biosorption of Cr(◉) from aqueous solution by chemically studies. Ecotoxicol. Environ. Saf.., 92: 199–205.

[6] Repo E., Warchol J. K., Bhatnagar A., Sillanpaa M. (2011) Heavy metals adsorption by novel EDTA-modified chitosan-silica hybrid materials. J. Colloid Interface Sci., 358: 261–267.

[7] Ren X., Chen C., Nagatsu M., Wang X. (2011) Carbon nanotubes as adsorbents in environmental pollution management: A review. Chem. Eng. J., 170: 395–410.

[8] Zhu J., Yang J., Deng B. (2010) Ethylenediamine-modified activated carbon for aqueous lead adsorption. Environ. Chem. Lett., 8: 277–282.

[9] Mellah A., Chegrouche S., Barkat M. (2006) The removal of uranium(VI) from aqueous solutions onto activated carbon kinetic and thermodynamic investigations. J. Colloid Interface Sci., 296: 434–441.

[10] Kumar S., Loganathan V. A., Gupta R. B., Barnett M. O. (2011) An assessment of U(VI) removal from groundwater using biochar produced from hydrothermal carbonization. J. Environ. Manage., 92: 2504–2512.

[11] Chen C., Li X., Zhao D., Tan X., Wang X. (2007) Adsorption kinetic, thermodynamic and desorption studies of Th(IV) on oxidized multi-wall carbon nanotubes. Colloid Surf. A: Physicochem. Eng. Asp., 302: 449–454.

[12] Haider S., Bukhari N., Park S. Y., Iqbal Y. (2011) Adsorption of bromo-phenol Blue from an aqueous solution onto thermally modified granular charcoal. Chem. Eng. Res. Des., 89: 23–28.

[13] Ricordel S., Taha S., Cisse I., Dorange G. (2001) Heavy metals removal by adsorption onto peanut husks carbon: Characterization, kinetic study and modeling. Sep. Purif. Technol., 24: 389–401.

[14] Tuzun I., Bayramoglu G., Yalcin E., Basaran G., Celik G., Arica M. Y. (2005) Equilibrium and kinetic studies on biosorption of Hg(II), Cd(II) and Pb(II) ions onto microalgae *Chlamydomonas Reinhardtii*. J. Environ. Manage., 77: 85–92.

[15] Sun Y. B., Yang S. T., Sheng G. D., Guo Z., Wang X. (2012) The removal of U(VI) from aqueous solution by oxidized multiwalled carbon nanotubes. J. Environ. Radioact., 105: 40–47.

[16] Kobya M., Demirbas E., Senturk E., Ince M. (2005) Adsorption of heavy metal ions from aqueous solutions by activated carbon prepared from apricot stone. Bioresour. Technol., 96: 1518–1521.

[17] Nadeem R., Nasir M. H., Hanif M. S. (2009) Pb(II) sorption by acidi-cally modified *Cicer arientinum* biomass. Chem. Eng. J., 150: 40–48.

[18] Duranoğlu D., Trochimczuk A. W., Beker U. (2012) Kinetics and thermodynamics of hexavalent chromium adsorption onto activated carbon derived from acrylonitrile–divinylbenzene copolymer. Chem. Eng. J., 187: 193–202.

[19] Bowman B. S. (2003) Applications of surfactant-modified zeolites to environmental remediation. Rev. Micropor. Mesopor. Mater., 61: 43–56.

[20] Xie J., Li C. J., Chi L. N., Wu D. Y. (2013) Chitosan modified zeolite as a versatile adsorbent for the removal of different pollutants from water. Fuel., 103: 480–485.

[21] Yu L., Gong J., Zeng C. F., Zhang L. X. (2013) Preparation of zeolite-A/chitosan hybrid composites and their bioactivities and antimicrobial activities. Mater. Sci. Eng. C., 33: 3652–3660.

[22] Wan Ngah W. S., Fatinathan S. (2010) Pb(II) biosorption using chitosan and chitosan derivatives beads: Equilibrium, ion exchange and mechanism studies. J. Environ. Sci., 22: 338–346.

[23] Soltani R. D., Khataee A. R., Safari M., Joo S. W. (2013) Preparation of bio-silica/chitosan nanocomposite for adsorption of a textile dye in aqueous solutions. Int. Biodeterior. Biodegradation., 85: 383–391.

[24] Aleksandra R. N., Sava J. V., Dušan G. A. (2013) Modification of chitosan by zeolite A and adsorption of Bezactive Orange 16 from aqueous solution. Composites: Part B., 53: 145–151.

[25] Gandhi M. R., Meenakshi S. (2012a) Preparation and characterization of La(III) encapsulated silica gel/chitosan composite and its metal uptake studies. J. Hazard. Mater., 203- 204: 29–37.

[26] Gandhi M. R., Meenakshi S. (2012b) Preparation and characterization of silica gel/chitosan composite for the removal of Cu(II) and Pb(II). Int. J. Biol. Macromol., 50: 650–657.

[27] Topallar H., Bayrak Y. (1998) The effect of temperature on the dynamic viscosity of acetone sunflower-seed oil mixtures. Turk. J. Chem., 22: 361–366.

[28] Sarin V., Pant K. K. (2006) Removal of chromium from industrial waste by using eucalyptus bark. Bioresour. Technol., 97: 15–20.

[29] Dahiya S., Tripathi R. M., Hegde A. G. (2008) Biosorption of lead and copper from aqueous solutions by pre-treated crab and arca shell biomass. Bioresour. Technol., 99: 179–187.

[30] Hall K. R., Eagleton L. C., Acrivos A., Vemeulen T. (1966) Pore and solid diffusion in fixed bed adsorption under constant pattern condition. Ind. Eng. Chem. Fund., 5: 213–223.

[31] Ho Y. S., McKay G. (1999) Pseudo-second order model for sorption processes. Process Biochem., 34: 451–465.

[32] Bhat S. V., Melo J. S., Chaugule B. B., D`Souza S. F. (2008) Biosorption characteristics of uranium from aqueous medium onto *Catenella repens*, a red alga. J. Hazard. Mater., 158: 628–635.

[33] Li Y. H., Xu C. L., Wei B. Q., Zhang X. F., Zheng M. X., Wu D. H., Ajayan P. M. (2002) Self-organized ribbons of aligned carbon nanotubes. Chem. Mater., 14: 483–485.

[34] Lu C. S., Chung Y. L., Chang K. F. (2005) Adsorption of trihalomethanes from water with carbon nanotube. Water Res., 39: 1183–1189.

[35] Shao D. D., Jiang Z. Q., Wang X. K. (2009) SDBS modified XC-72 carbon for the removal of Pb(II) from aqueous solutions. Plasma Process. Polym., 7: 552–560.

[36] Shao D. D., Jiang Z. Q., Wang X. K., Li J. X., Meng Y. D. (2009) Plasma induced grafting carboxymethyl cellulose on multiwalled carbon nanotubes for the removal of UO_2^{2+} from aqueous solution. J. Phys. Chem., B113: 860–864.

[37] Harmsen K. F., Haan A. M. (1980) Occurrence and behavior of uranium and thorium in soil and water. Neth. J. Agric. Sci., 28: 40–62.

[38] Hsi C-K. D., Langmuir D. (1985) Adsorption of uranyl onto ferric oxyhydroxides: Application of the surface complexation site-binding model. Geochim. Cosmochim. Acta., 49: 1931–1941.

[39] Katsoyiannis I. A. (2007) Carbonate effects and pH-dependence of uranium sorption onto bacteriogenic iron oxides: Kinetic and equilibrium studies. J. Hazard. Mater., B139: 31–37.

[40] Duquènea L., Tack F., Meers E., Baeten J., Wannijna J., Vandenhovem H. (2008) Effect of biodegradable amendments on uranium solubility in contaminated soils. Sci. Total Environ., 391: 26–33.

[41] Chingombe P., Saha B., Wakeman R. J. (2005) Surface modification and characterisation of a coal-based activated carbon. Carbon, 43: 3132–3143.

[42] Wang H., Zhou A., Peng F., Yu H., Chen L. (2007) Adsorption characteristic of acidified carbon nanotubes for heavy metal Pb(II) in aqueous solution. Mater. Sci. Eng., 466: 201–206.

[43] Ahmad A., Rafatullah M., Sulaiman O., Ibrahim M. H., Chii Y. Y., Siddique B. M. (2009) Removal of Cu (II) and Pb (II) ions from aqueous solutions by adsorption on sawdust of Meranti wood. Desalination, 247: 636–646.

[44] Ricordel S., Taha S., Cisse I., Dorange G. (2001) Heavy metals removal by adsorption onto peanut husks carbon: Characterization, kinetic study and modeling. Sep. Purif. Technol., 24: 389–401.

[45] Li Y. H., Wang S. G., Luan Z. K., Ding J., Xu C. L., Wu D. H. (2003) Adsorption of cadmium(II) from aqueous solution by surface oxidized carbon nanotubes. Carbon, 41: 1057–1062.

Chapter 5

Microalgal Biorefineries

Eduardo Jacob-Lopes, Luis Guillermo Ramírez Mérida,
Maria Isabel Queiroz and Leila Q. Zepka

Additional information is available at the end of the chapter

http://dx.doi.org/10.5772/59969

1. Introduction

The world has been confronted with a food and energy crisis due to accelerated global population growth and the depletion of finite fossil fuel resources. The increase in nutritional problems along with rising fuel demands and environmental problems have necessitated the search for nutritional supplements and sustainable sources of energy. Currently, fossil fuel resources are not regarded as sustainable and their continued consumption is raising serious ecological, economic and environmental questions. However, while we move towards alternative sources of energy, there remains a need to replace fossil fuels with high energy density fuels. A highly contentious issue of great concern is the argument that emissions of carbon dioxide (CO_2) from fossil fuel use, especially from coal combustion, are responsible for global climate change. As a result of studies during the past five decades, and most notably from the last 20 years, emissions of CO_2 have become an important issue with respect to global climate change because atmospheric CO_2 concentrations increased significantly in the last century and have continued to rise at an increasing rate [1].

The United Nations Kyoto Protocol of 1997 established regulations designed to control emissions of air pollutants with the objective of reducing greenhouse gases to the level of emissions in 1990, and more than 170 countries have ratified the protocol [2].

Various CO_2 sequestration techniques have been developed and the various technologies for CO_2 capture and storage need to be evaluated from the point of view of obtaining carbon credits, aimed at stabilizing emissions of the pollutant [2]. Of these techniques, CO_2 capture by photosynthetic organisms such as microalgae shows good potential in view of the economic advantages it presents, rate of CO_2 capture, and the speed with which the technology can be introduced to the industrial community.

Microalgae have great potential in generating energy from biotechnological processes using renewable sources and without compromising food security and agriculture. Microalgae have been of major interest in biofuel production as well as in the feed, chemical and pharmaceutical sectors [3]. Depending on the species and growth conditions, microalgae can be selected to produce a wide variety and abundance of lipids, proteins, carbohydrates, and feedstocks important for biofuel and production of nutraceuticals [3].

The rapid growth rate coupled with high productivity from a small area means that the production of microalgal biomass has a promising future [4].

Several investigations into the use of microalgae to obtain bioproducts have been successfully conducted. Upstream processing (USP) and downstream processing (DSP) are stages found in the processes of microalgae biorefineries (Fig. 1). USP involve four important areas: (i) microalgae strain, (ii) CO_2 supply, (iii) nutrient source <nitrogen/phosphorus> and (iv) source of illumination [5].

Photobioreactors used for the culture of microalgae are of two basic designs — open or closed systems. Amongst the different types of open system design, the most popular is the raceway pond, while popular closed systems include flat-plate, vertical tubular, horizontal tubular and hybrid type photobioreactors. Growth of microalgae in photobioreactors occurs due to the use of CO_2 rich gas as a means of mixing, as well as being a source of carbon. Generally, in this type of reactor, the agitation, mass transfer, efficient provision of light, removal of photosynthetically generated oxygen, understanding of hydrodynamic aspects, and scalable photobioreactor technology are aspects that should be taken into account to achieve good yields [6,7].

Conventional DSP includes all unit processes that follow the process taking place within the photobioreactor. They involve biomass harvesting and biorefinery techniques which facilitate the integration of the biomass conversion processes and equipment for the production of several fractions of interest through the use of mild separation technology. Biorefining involves assessment and use of different technologies to obtain different types of bioproducts from biomass, which can be marketed and used to solve specific problems in many different areas. Finally, there must be the safe and inexpensive disposal of all waste products generated during the process. Therefore, a portion of the residual biomass can go to an anaerobic digester to generate biogas, and the rest can be used as nutrients to feed the photobioreactor again.

As such, the aim of this chapter is to present an overview of the potential uses of the technology in the transformation of carbon dioxide into biomolecules, and to describe the processes involved in the biological conversion of CO_2 in photobioreactors as well as biorefinery techniques suitable for the treatment of microalgal biomass and the production of biomolecules.

2. Carbon dioxide emissions and mitigation

Climate change occurs mainly due to increased levels of CO_2 in the atmosphere. During the twentieth century an increase in CO_2 concentration of 30% was observed. This rate of increase

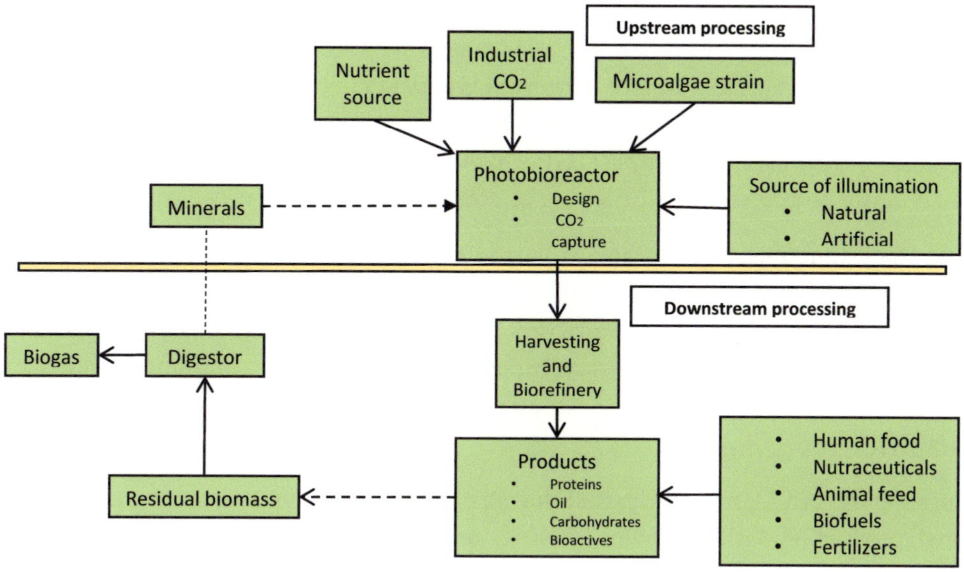

Figure 1. Outline of the formation process of microalgal biomass and bioproducts.

will lead to an increase in CO_2 levels of 49% by the end of this century [8]. CO_2 emissions from fossil fuel combustion saw an increase of 41% between 1990 and 2008 [9]. Three potential sources of CO_2 can be found: stationary, mobile and natural. CO_2 is the primary greenhouse gas emitted through human activities. Stationary sources contribute the highest percentage of CO_2 emissions of these, and are anthropogenic in origin, such as from industrial or domestic processes. The industrial processes contributing to increasing atmospheric CO_2 concentrations consist of hydrogen and ammonia production plants, power stations, cement companies, ethanol companies, and chemical factories. Flue gases from power plants are responsible for more than 8% of the world's total CO_2 emissions. Mobile sources are those from transport, while natural sources include volcanoes and elements of human or animal decomposition [10].

A number of research and development efforts have been directed at reducing CO_2 emissions. Many of these studies use different microalgae strains or new photobioreactors with geometric configurations that may be a fundamental step forward for the consolidation of this technology [11-14].

Biological methods for CO_2 mitigation can be carried out by photosynthetic microorganisms such as microalgae and plants, the latter with an estimate for CO_2 capture of only 3–6% of fossil fuel emissions [15]. Biomitigation using microalgae as a method of decreasing CO_2 emissions by fixing CO_2 through photosynthesis is considered one of the most effective. Generally, microalgae have higher growth rates, a higher CO_2 fixation efficiency and a larger production of biomass enabling the subsequent development of quantities of bioproducts with high added value [16].

However, much the flue gas of industrial origin, in addition to contributing to CO_2 emissions, produces other compounds, including oxygen (O_2), water vapour, carbon monoxide (CO), nitrogen oxides (NOx), sulfur oxides (SOx), hydrochloric acid, heavy metals, and particulate matter [17]. These compounds have a toxic and/or inhibitory effect on microalgal growth. Tolerance of microalgae to elements of flue gas is dependent on the strain. NOx present in flue gas can be taken up and could become an alternative nitrogen source for the growth of microalgae growth. The main impact is due to SOx which reacts with water to form sulfurous acids. This can be prevented by buffering or active pH control [18].

Besides high amounts of NOx and SOx, the high temperature influences the growth of microalgae. Industrial plants discharge flue gas at a temperature between 70–120 °C [19]. Therefore, to complete the CO_2 capture process, it is necessary either to install a post-cooling system, or to use thermophilic species [18]. Additionally, oxidant compounds found in flue gas can cause damage to proteins and pigments, and compromise the integrity of cell membranes [20].

Besides the use of CO_2 from industrial flue gas, other alternative sources are known such as ethanol production facilities, winegrowing, ammonia and hydrogen production or gas-processing plants. The capture of CO_2 from the fermentation process is relatively simple and cheap due to the higher state of purity in which it is present [21].

Algae companies	Country	Description	Ref
AFS BioOil Co.	San Francisco, USA	Founded in 2010. This company uses algae that are fed by nutrients recovered from wastewater treatment plants, electricity generation, and sunlight. Biodiesel, the main product produced in these biorefineries is cost competitive with petroleum products.	[22]
AFS Biofarm™	San Francisco, USA	Uses CO_2 sequestered from industrial facilities and power plants for conversion into renewable fuels and other valuable products such as food additives.	[23]
Seambiotic Ltd.	Tel Aviv, Israel	Founded in 2003. The company aims to develop microalgae biomass for the production of food additives and biofuel using flue gas from coal burning power stations.	[24]
Aeon Biogroup	Chile	This company develops biomass production methods with CO_2 capture from winegrowing for production of oil, nutraceuticals, food additives and biochemical compounds.	[25]
IGV Biotech	Nuthetal, Germany	Founded in 1960. IGV Biotech develops microalgae biotechnology processes for the production of several products such as food, pharmaceuticals and chemicals. This company uses advanced technology for the cultivation of photosynthetic microorganisms and CO_2 capture.	[26]
Algenol Biofuels	Florida, USA	Founded in 2006. The company uses CO_2 and seawater as a culture medium for bioethanol. Nitrogen fixing technology is used to reduce production costs of fertilizers by cyanobacteria.	[27]

Table 1. Global companies with CO_2 sequestering technology for algae culture.

Table 1 shows some global companies which employ CO_2 sequestering technologies for the production of biofuel and/or bioproducts from algal cultures. Furthermore, many companies and research centers worldwide are investigating the upstream and/or downstream process.

3. Microalgae strains and photosynthetic metabolism

Microalgae are fast growing photosynthetic microorganisms that produce valuable compounds, are easy to harvest, exhibit a unicellular or simple multicellular structure, and a large surface-to-volume body ratio. Eukaryotic microalgae such as green algae (*Chlorophyta*) and diatoms (*Bacillariophyta*) as well as prokaryotes like cyanobacteria species (*Cyanophyceae*) use oxygenic photosynthesis to fix CO_2 like macroalgae and plants [18].

CO_2 fixation, biomass production and bioproduct diversity vary with microalgal species, although the data may not be strictly comparable as the microalgae may have different biological behavior or may have been cultured in different conditions. The general chemical composition of different microalgal species varies, with some species having greater potential for the production of certain bioproducts [28]. Microalgae have a varied biochemical profile (Table 2). The high protein content of microalgae species is notable. These proteins, mainly amino acids, provide nutritional elements that can meet food requirements in humans and animals.

Consequently, a successful and economically viable microalgae industry producing bioproducts mainly depends on the selection of appropriate microalgae strains.

Microalgae comprise a diversity of species characterized by a variety of phenotypes dependent on their pigments and cell structure. Chlorophyll-*a* and phycobiliproteins may be present and are involved in harvesting light energy for photosynthesis. They are, therefore, a good choice for the generation of biomass. In addition to photosynthesis, some species show an ability to adapt to different environments and metabolisms such as respiration and nitrogen fixation, chromatic adaptation and the ability to form symbiotic associations with yeast, fungi, bacteria and plants [31]. The part of the photosynthetic process in which CO_2 is converted into carbohydrates is catalyzed by the carboxylase activity of the enzyme ribulose 1,5-bisphosphate carboxylase/oxygenase (RuBisCO), this step is called the Calvin cycle [32]. The Calvin cycle is the metabolic mechanism for CO_2 fixation in microalgae and the process comprises three phases: carboxylation, reduction and regeneration. The photosystem II (PSII) complex is the starting point of photosynthesis, where via the electron transport chain, an electron is transferred to cytochrome b6f and PSI. A proton-motive force is created due to the pumping of electrons in opposite directions, creating a difference in charge across the membrane. This is used for ATP synthesis and the formation of ferredoxin and NADPH. The electron is donated by water and oxygen is formed as a waste product [33,34]. To generate one molecule of carbohydrate (CH_2O), O_2 and H_2, at least eight (8) photons are needed. The mean energy content for photosynthetically active radiation (PAR) photons is close to 220 kJ/mol and the total potential light energy captured by photosynthesis is 1744 kJ/mol of CH_2O. The theoretical

Microalgae species	Protein	lipid	Carbohydrate	Nucleic acid	Ref.
Anabaena cylindrical	43–56	4–7	25–30	–	[28]
Aphanothece microscopica Nägeli	41–49	8–9	13–18	3–4	[29]
Arthrospira maxima	60–71	6–7	13–16	–	[28]
Chlamydomonas rheinhardii	48	21	17	–	[28]
Chlorella pyrenoidosa	57	2	26	4–5	[28]
Chlorella vulgaris	51–58	14–22	12–17	4–5	[28]
Dunaliella bioculata	49	8	4	–	[30]
Dunaliella salina	57	6	32	–	[30]
Euglena gracilis	39–61	14–20	14–18	–	[28]
Porphyridium cruentum	28–39	9–14	40–57	–	[30]
Prymnesium parvum	28–45	22–38	25–33	1–2	[30]
Scenedesmus dimorphus	8–18	16–40	21–52	3–6	[28]
Scenedesmus obliquus	50–56	12–14	10–17	3–6	[28]
Spirogyra sp.	6–20	11–21	33–64	–	[28]
Spirulina maxima	60–71	6–7	13–16	2–5	[28]
Spirulina platensis	46–63	4–9	8–14	2–5	[28]
Synechoccus sp.	63	11	15	5	[30]

Table 2. Chemical composition of different microalgae expressed on a dry matter basis (%).

maximum efficiency for the conversion of light to ATP is approximately 27%. However, only 42.3% of PAR can be utilized [35]. Furthermore, light intensity and light quality also play a key role in the growth rate of the cell [36].Inorganic and organic carbon represent one of the main sources of nutrition for microalgae. Different microalgae species can be maintained at various concentrations of CO_2 due to a mechanism called the carbon concentration mechanism (CCM), which accumulates inorganic carbon, concentrating it in the CO_2 in RuBisCO [37].

Microalgae can assimilate carbon through three routes: (i) direct absorption of CO_2 by the cell membrane; (ii) the use of bicarbonate by inducing the enzyme carbonic anhydrase and (iii) transport of HCO_3^- by the cell membrane. The enzyme carbonic anhydrase catalyzes the reaction converting HCO_3^- into CO_2, moreover RuBisCO uses CO_2 as the substrate on which it forms phosphoglycerate. Limitations to CO_2 production can occur, slowing the rate of reaction. Thus, carbonic anhydrase is a very efficient enzyme that can generate high concentrations of CO_2 [38–40].

4. Requirements of microalgae biorefineries

In addition to CO_2, other nutrients such as phosphorus, nitrogen and trace metals are important for the production of microalgae. These provides the necessary conditions for microalgae to carry out the metabolic reactions necessary for growth and so generate biomass or primary metabolites [41]. Most microalgae species can utilize inorganic and organic nitrogen sources.

Ammonium salts, ammonium sulfate, diammonium hydrogen phosphate and ammonia are supplied as inorganic nitrogen sources, and urea is supplied as an organic nitrogen source [42–43]. Phosphorus is supplied as hydrogen phosphate and dihydric phosphate in small amounts. Normally, sufficient quantities of minerals such as cobalt, copper, iron, molybdenum, manganese and zinc are present in the water supply, or they may be added as specific salts [44]. An adequate supply of nutrients is a prerequisite for high production rates. The introduction of certain nutrient stresses may affect the biochemistry. For example, nitrogen stress is important for carotenogenesis in *Dunaliella salina* [45] and increased lipid production in *Chlorella vulgaris* [46].

The most important factor in CO_2 fixation and microalgae biomass production is light intensity. Light sources can be divided into natural sunlight, which is applied in both open and closed cultivation, and artificial cold light that is mainly applied in closed cultivation. Various studies have been performed on the use of artificial light. Many lamps are available commercially such as light emitting diodes (LED), fluorescent tubes, halogen, tungsten, and high intensity discharge lamps (HID), and optical fibers. The investment costs, shelf life and stability of light intensity are important factors to consider when choosing a lighting source [47]. Recommended light sources for microalgae cultivation include the following: In laboratory research fluorescent tubes exhibiting a PAR efficiency of 1.25 μmol-ph s^{-1} W^{-1} are used, while HID with a PAR efficiency of 1.87 μmol-ph $s^{-1}W^{-1}$ are the most commonly employed in horticulture, along with LED lamps with a PAR efficiency of 1.91 μmol-ph$s^{-1}W^{-1}$ [47]. The same type of lamp can emit different wavelengths, with blue LED and red LED having an adsorption at around 450–470 nm and 645–665 nm respectively [48]. Wang et al. [49] found the highest biomass yield using red LED in *Spirulina platensis* cultivation [49]. An analysis of Table 3 demonstrates a variety of possible sources of illumination for use in microalgae production.

The rate of photosynthesis is proportional to light intensity. When irradiance is increased, microalgal growth rate accelerates, but exposure of cells to long periods of high light intensity causes photoinhibition. Microalgae can only utilize the energy available in the 400–700 nm wavelength range, represented by PAR [50]. Moreover, several studies reported that the optimal wavelength varied from species to species [51]. Light intensities of 100 and 200 $\mu E/m^2/s$ are frequently used [52].

Besides light intensity, it was found that light–dark cycles could also significantly influence microalgal growth by avoiding sustained exposure to high photon flux density and providing dark time for microalgae to repair photo-induced damage [53].

The light–dark periods for most microalgal cultivation are 24h:0h,16h:8h and 12h:12h. This varies with microalgal species. Experiments carried out with *Aphanothece microscopica Nägeli* in photoperiods of (22:2), showed that growth rate is not conditional on incident lighting over 22 h. This provides evidence that *Aphanothece microscopica Nägeli* have the possibility of storing energy for their biochemical processes, without affecting the rate of photosynthetic metabolism [54].

On the one hand, the light intensity of natural sunlight is cheaper, but the light cycle depends on weather and latitude, which often preclude higher biomass production. At the same time, artificial illumination is generally expensive, but the control of light intensity afforded is excellent, allowing for greater flexibility and constant biomass production [55].

Light source	Commentary	Electrical consumption	Stability	Investment cost	Weather durability	Ref.
Conventional lamp (halogen, tungsten, fluorescent)	High productivity of biomass, large area lighting, generation of high temperature	High	High	Low	Moderate	[47]
LED lamp	Low heat generation, Greater resistance to on/off cycles	Low	High	Low	High	[47,56]
HID lamp	Generation of high temperature, high efficiency. However, losses from trapped light in protective covers and lenses, inefficient ballasts	Low	High	Low	Moderate	[47]
Optical fiber excited by lamps	Small space requirements for installation, good light distribution, uniformity of illumination, low risk of contamination	High	Moderate	High	High	[57]
Natural sunlight	Variable biomass productivity depending on weather conditions, good lighting area, economic and adequate light distribution	Absent	Low	Low	High	[55]

Table 3. Characteristics and power consumption for different sources of artificial light

A photobioreactor is a device consisting of an illuminated culture vessel designed for the controlled bioconversion of CO_2 into biomass and bioproducts. The two basic types of photobioreactors used for the large scale culture of microalgae are open or closed systems. Open systems can be built more easily, are more economical and relatively simple to control in relation to closed systems. Most open systems are natural lakes or open ponds. Two types of open systems are known: (i) circular ponds stirred with a rotating arm and (ii) raceway ponds, which are shallow artificial ponds divided into a rectangular grid with paddle wheels for culture mixing. Raceway ponds are the most popular open system design [58].

The reactor surface is illuminated with natural light and the intensity of illumination affects the microalgae culture. The depth of this type of reactor may not exceed 35–40 cm so that it does not prevent the passage of light to the bottom of reactor. The reactor performance declines with increasing depth due to the fact that diminishing amounts of light energy are available.

Moreover, the use of an open system for the sole purpose of CO_2 sequestration is mitigated by the very low residence time of gas in the culture, which therefore offers a short time in which the fixing of CO_2 from flue gases by the microalgae can occur. Open systems produce low yields of products with high added value due to contamination problems [55]. Closed systems support high yields of microalgae biomass and they have certain advantages with regard to minimizing contamination, allowing axenic microalgal cultivation, providing a control system for various parameters such as pH, temperature, light, and CO_2 concentration. They also reduce CO_2 losses, prevent water evaporation, allow for a greater control of biomass growth, and permit the production of complex biomolecules. Closed systems are currently being assessed for microalgae cultivation in configurations such as flat-plate, vertical tubular, horizontal tubular and hybrid type photobioreactors [59].

Flat-plate and tubular photobioreactors are the commonest types used for cultivation of microalgae in the laboratory and on a pilot scale. These photobioreactors are based on the same principles of a large surface-to-volume (S/V) ratio, optimal use of CO_2 and suitable mixing. Tubular photobioreactors (airlift or bubble column) seem the most suitable for CO_2 sequestration due to their homogeneous mixture, greater gas transfer, smaller hydrodynamic stress, ease of construct and high productive output. Flat-plate photobioreactors are very expensive to build, which makes them unfeasible for industrial use. A hybrid photobioreactor is a combination of at least two types of different photobioreactor. Usually, integrating a horizontal tubular photobioreactor with a vertical tubular photobioreactor will compensate for the drawbacks in scale-up and the enhanced S/V ratio of vertical tubular photobioreactors. There are many configurations that have been studied, producing good results [55].

Photobioreactor development is perhaps one of the major steps that should be undertaken for the efficient large-scale cultivation of microalgae and bioproduct formation. Shape considerations must be taken into account when installing a system to produce bioproducts. Closed reactors are best for production of compounds of high added value.

There is a complex CO_2 transfer process in a photobioreactor. In the gas aerating method, mass transfer performance and biochemical reaction rate depend of the type and size of the photobioreactor, the range of operational conditions, the influence of physicochemical properties on hydrodynamics due to the high viscosity of the liquid, its rheological behavior, the measuring method used, bubble size, gas hold-up, the gas/liquid contact area, and CO_2 concentration and gas/liquid ratio [55].

In terms of solubility, oxygen is less soluble in water than CO_2. However, both gases are poorly soluble in aqueous solution so there is a need for the provision of these elements throughout the process. CO_2 bubbling in solution alone does not produce complete dissolution, since a fraction of the CO_2 injected is lost in the gas outlet. The chemical reactivity of CO_2 in the water forms H_2CO_3. The pH decreases with increasing insolubility and carbonic acid formation. The H_2CO_3 dissociates to HCO_3^- and CO_3^{2-}. Consequently, the total inorganic carbon concentration is represented by the totality of the compounds CO_3^{2-}, HCO_3^- and CO_2 [4].

The volumetric mass transfer coefficient (K_La) is the property of the photobioreactor that determines the appropriate conditions that will ensure cell growth in the reactor. K_La represents a function of microalgal characteristics and operating conditions. Efficiency of CO_2 transfer is necessary to increase the K_La of CO_2 allowing for improved transfer of gas to the liquid phase. Photobioreactors require an efficient CO_2 transfer system [4].

During microalgae growth in the photobioreactor, the accumulation of O_2 can occur. The water dissociation activity of PSII is responsible for the oxygen produced during photosynthesis. The increase of O_2 in the culture medium is a hard problem to solve. The level of dissolved O_2 in the culture medium causes photoinhibition reducing photosynthetic efficiency. Accumulation of O_2 becomes a complicated problem in a closed photobioreactor when the reactor configuration does not provide an interface between the culture and the surrounding atmosphere, in contrast to horizontal tubular reactors or a flat panel configuration. The solutions proposed to date rely on the installation of a degasser. In photobioreactors, degassing is only necessary when O_2 production (due to basal activity of PSII) is higher than respiration [60].

5. Microalgal biomass harvesting technologies

When the biochemical process in the photobioreactor have finished, the upstream processing ends and gives way to downstream processing and harvesting of the biomass and refining of the bioproducts in the biorefinery. As a result, most of the production costs in microalgae biorefineries are influenced by DSP. The microalgal harvest and reduction of its water content does not depend on a single method. Efficient and profitable harvesting methods are required to process the biomass and bioproducts economically [61]. The microalgae recovery techniques represent between 20–30% of total production costs. Of harvesting techniques, the most commonly used are flocculation, filtration, centrifugation, gravity sedimentation and flotation [62]. Some factors influencing the choice of harvesting technique are morphology, density and size of the microalgae, as well as the type and quality of product to be obtained [35].

Flocculation is a process in which particles are dispersed from the medium by using chemicals to aggregate the microalgal cells. Flocculants stimulate flocculation by causing colloids and other suspended particles in solution to form flocs. Chemical flocculants regularly used to harvest microalgae include ferric chloride, aluminum sulfate, alum, ferric sulfate, polyferric sulfate, as well as cationic polymers (polyelectrolytes), and organic flocculants (chitosan). Researchers have developed a process of cell autoflocculation, through the adjustment of pH in the microalgae culture [63].

Filtration operated under pressure or in a vacuum is satisfactory for recovering relatively large (>70mm) and/or filamentous microalgae, but is less effective in separating microalgae species with dimensions close to those of prokaryotes. Membrane microfiltration and ultrafiltration processes may be an option for the recovery of microalgae biomass under 30 mm [63]. Petrusevski [64] recovered 70–89% of microalgae using cross flow filtration with the advantage of maintaining the integrity of the microalgae biomass. On the other hand, use of a chamber-membrane filter press could achieve a concentration factor of 245 times the original concentration for *Coelastrum proboscideum* and produce a sludge with 27% solids [65]. Filtration is an expensive process due to membrane exchange and pumping. At larger scales of production

(>20 m^3 per day) other methods can be cheaper. For the processing of small volumes (e.g., < 2 m^3 per day) it can be more cost effective when compared with centrifugation [66].

Centrifugation is a methodology that includes the use of centrifugal acceleration for the sedimentation of microalgae in heterogeneous mixtures. The size and density of the structure determines the centrifugal separation of element. The supernatant is a liquid located in the upper layer of the centrifuge tube and the microalgae concentrate is represented by the remnant solid. The fast and intensive process depends on the sedimentation of cells for biomass recovery, as well as the amount of time of the cell suspension is in the centrifuge, and settling depth [63]. This method can lead to cell injury due to the gravitational and shear forces encountered [67]. Many researchers have recommended this method for the reliable recovery of microalgae [68–70]. Recovery by centrifugation is an efficient method when used with small volumes of fluid and high energy consumption. The drawbacks of centrifugation are the high initial investment costs, the noise generated during operation, and the cost of the electricity used [71].

Gravitational sedimentation is widely used to separate microalgae in aqueous solution and for wastewater treatment. The sedimentation rates of microalgae are influenced by the rate of sedimentation of solids and are determined by the density and area of the microalgae cells [35]. Gravitational sedimentation, preceded by flocculation, is one of the most widely-used techniques for the harvesting of microalgae biomass. Disadvantages of the method are that is very slow (0.1 to 2.6 m h^{-1}) and the biomass may suffer decomposition under conditions of high temperature. Furthermore, the technique is suited for use with large microalgae or those with a filamentous morphology [72].

Flotation methods are based on the binding of microalgae cells using air bubbles. The resulting flocs rise to the surface of the liquid and are recovered by either physical or chemical procedures. Particles as small as 500 μm can be recovered by flotation. Some strains have gas vacuoles and float at the surface of the water [72]. The incorporation of air bubbles depends on several aspects such as the contact angle of air, solid, and aqueous phases. According to the method of bubble production, flotation techniques can be divided into dissolved air flotation (DAF), dispersed flotation and electrolytic flotation. DAF is the most used method in the treatment of industrial wastewater. Microalgae cells are recovered by dissolving air in the water under pressure and are then released into a reservoir at atmospheric pressure thus producing small bubbles. Chemical flocculation has been used with DAF to separate microalgae [73]. Garg et al. [74] evaluated the effects of froth flotation in different microalgae strains and found that *Chlorella* sp. showed a good response of floatability due to its high hydrophobicity. This method represents a promising choice for the industrial scale harvesting of microalgae and represents a very versatile technique for the separation of small particles. Microalgae with low surface hydrophobicity are difficult to harvest by flotation separation. Surface hydrophobicity and bubble size are the key factors affecting algae flotation, and a stepwise optimization can lead to effective separation by flotation of difficult-to-harvest microalgae [64]. Although flotation has been widely used by researchers as a harvesting method, there are feasibility and economic limitations.

6. Biorefinery

Biorefinery comprises a number of specialized methods used to extract the most out of primary and/or secondary metabolic products. Microalgae biorefineries must use methods and technology for isolating compounds and obtaining principal constituents from biomass, without damaging one or more of the product fractions, thereby adding value to the bioproduct formed [75].

The main focus when obtaining the products should be the dehydration or drying of the biomass when there is a requirement for its immediate use, otherwise the harvested biomass suspension must be processed rapidly. Methods used for drying microalgae include lyophilization, spray drying, drum drying and sun drying. The next step is cell disruption as some target products are intracellular and therefore cell disruption is required in order to release the products and ready them for extraction. Several methods can be used depending on the metabolites of interest [76].

The product fractions obtained from microalgae can be transformed into high-value molecules, antioxidants, anti-inflammatories, natural pigments, biofuels, and food supplements for human and animal feed. Microalgae biorefinery is, therefore, a process of great industrial impact and must be undertaken properly (Figure 2).

Methods of drying microalgal biomass are used with the purpose of increasing the longevity of cells. Drying methods may include sun drying, lyophilization, drum drying, spray drying, and fluidized bed drying. Sun drying is cheap but is very slow. Spray drying is the method chosen for obtaining bioproducts with a high added value, though this procedure is not recommended for extraction of microalgae pigments as it may affect the pigments' molecular structure. Lyophilization is widely used in scientific research procedures, but is very expensive for use on a large scale. It is useful with respect to some enzymes and pharmaceuticals. This method eliminates thermal and osmotic damage and preserves the cell constituents microalgae [77].

By comparing different drying techniques (sun drying, lyophilization) for the effective extraction of lipids from *Scenedesmus* sp. grown in a raceway reactor Guldhe et al. [77] showed that drying methods are critical for effective downstream processing in the synthesis of microalgal biodiesel. No statistically significant difference was found in the drying methods used for the extraction of lipids

The extraction of intracellular components requires the breakdown of the microalgal cell wall. Various disruption methods involving chemical treatments (solvents, acids), mechanical treatments (ultrasound, high-pressure homogenization, bead beating and blending), autoclaving, freezing–thawing sequences and supercritical fluids have been used. Some of the extraction and fractionation techniques will be described briefly below. Microalgae biorefineries seek to apply these methods at an industrial scale and at low cost. Their high functionality with low concentrated streams is advantageous. Choosing the most appropriate method depends of various biological factors and the energy required. The integration of cell disrup-

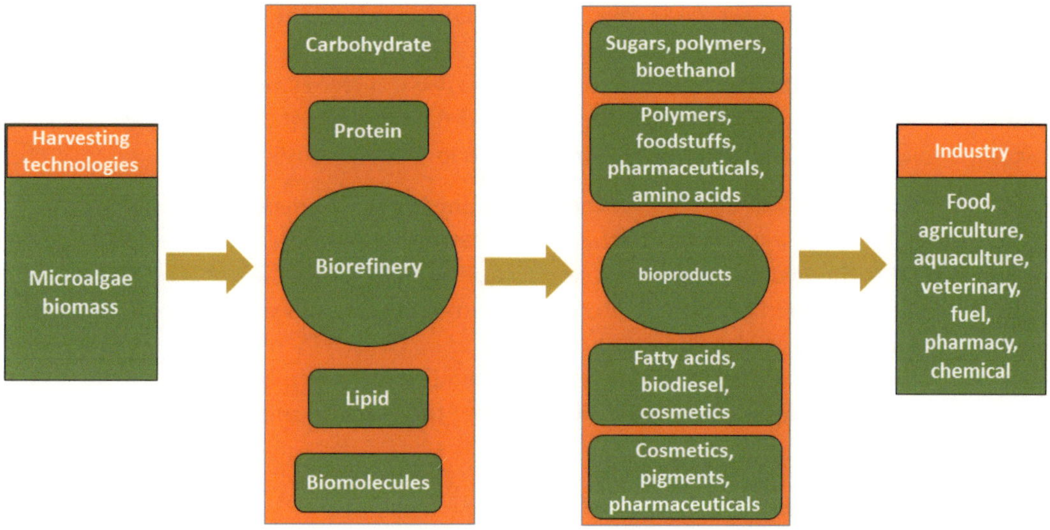

Figure 2. General outline of microalgae biorefinery.

tion into downstream processing has to be easy and should not have a negative impact on subsequent processing steps [78].

Pulsed electric field procedure can be a promising alternative to conventional cell disintegration methods. The procedure is based on the dosing of short electrical pulses in high intensity electric fields. These alter the structure of the cell membrane, which as a result loses its barrier function and becomes permeable — a phenomenon often referred to as electroporation [79].

Goettel et al. [78] evaluated the application of the pulsed electric field procedure for the disruption of *Auxenochlorella protothecoides* cells. For all pulse parameters applied, there was evidence that cell disintegration resulted in the release of soluble intracellular matter into the suspension. The efficiency of cell disruption improved with increasing treatment energy, whereas the field strength had no major influence. Thus, the investigation proposed the use of the pulsed electric field procedure of cell disruption and selective two-step extraction. As an initial step, the pulsed electric field procedure allows separation of water soluble intracellular substances. In a subsequent step, lipids can be very efficiently extracted by solvents.

The ultrasound method is based on the incorporation of high frequency sound waves in microalgae cells so that pressure variation can disrupt the cell wall [80].

Ehimen et al. [81] successfully used ultrasound to improve methods of oil extraction in *Chlorella* biomass samples. The results of the study showed that it is feasible to reduce volumes of methanol used in the trans-esterification process. The combination of an ultrasonic process and solvent use demonstrated the potential for recovery of greater yields of fatty acid methyl esters (FAME) for biodiesel production.

Enzymatic degradation is another method of cell wall disruption. This method is used on a laboratory scale since high costs limits its use on a larger, industrial scale. The advantage is that the enzyme may be inactivated, removed, recovered and reused. It has a high specificity without interfering in the recovery of bioproducts. Enzymatic disruption of microalgae cell walls can be performed with a mixture of β-glucanases and lysozyme. Studies of enzymatic hydrolysis of *Chlorella* cell walls have demonstrated the high specificity of the disruption so that mechanical degradation can be performed with low energy costs [82].

Chemical treatments using acid are performed by immersing diluted microalgae biomass in strong acid followed by a strong base, at high temperature for a specific time. One disadvantage of the method is the toxicity of the acid, and as a result the method is not widely used [83]. Sathish and Sims [83] demonstrated a method of extracting transesterifiable lipids using acid and base hydrolysis for *Chlorella* sp. and *Scenedesmus* sp. with 84% of moisture. On average, 60% of lipids were extracted and converted to biodiesel by transesterification. This was achieved without drying the recovered biomass and the use of a smaller volume of organic solvent was evident.

The choice of cell disruption method is dependent on the bioproduct, the strain of microalgae used, and the costs and efficiency of the process

7. Extraction and purification of microalgae metabolites

Solvent extraction systems are extensively used to extract microalgae metabolites from processed biomass. Solvents such as ethanol, chloroform, diethyl ether, hexane and methanol are commonly used. These can extract carbohydrates, amino acids, salts, hydrophobic proteins, lipids and pigments. The disadvantages of solvent extraction are that: (i) the process requires high capital investments; (ii) the energy requirements are high; (iii) the solvent is highly flammable; and (iv) the difficulty of recovering the solvent [84].

Different process for the extraction of fatty acid from *Aphanothece microscopica Nageli*, *Phaeodactylum tricornutum*, *Isochrysis galbana* have been described [29,85]. Extraction using aqueous buffers is employed to obtain phycobiliproteins from *P. cruentum* and lutein from three *Chlorella* species [86].

The regeneration of solvent for subsequent operation is difficult, further decreasing the efficiency of extraction. A method that can recover the solvent for reuse would be ideal from an economic point of view. This phase splitting could be induced by changing the nature of the solvent [87].

Du et al [87] studied the extraction of oil from *Desmodesmus* sp. by CO_2-switchable solvents. In this research, the secondary amines dipropylamine and ethylbutylamine were able to extract lipid from a liquid medium without damaging microalgae cells. These solvents allow the process of quick and efficient lipid extraction in the presence of water induced by the presence of ambient temperatures and atmospheric CO_2. These solvent systems provide a potential for reuse and recovery leading to decreased costs and provide an efficient method of microalgae lipid extraction [87]. Properties of the cell membrane are of great importance in the solvent extraction process. Therefore, disruption of the cell wall is critical [88].

Crude extracts are generally filtered and purified by several chromatographic methods in order to obtain the metabolite of interest. In choosing a chromatographic technique certain considerations should kept in mind. These include molecular weight, isoelectric point, hydrophobicity and biological affinity. Supercritical fluid extraction has been shown to be an efficient technique for extracting carotenoids from microalgae *Scenedesmus* sp. [89] and fatty acids, and of the three microalgae strains evaluated, *S. obliquus* is the best source of α-linolenic acid [90].

Some other chromatographic methods included reverse phase chromatography, silica gel adsorption chromatography, and ion exchange chromatography (for proteins). Chromatographic techniques are usually employed for higher-value products. An economical evaluation could be useful to help calculate the optimum conditions for industrial applications [91].

8. Potential uses for bioproducts obtained from microalgal biorefineries

Microalgae have massive potential to produce biomolecules due to the low cost of energy and nutrient sources used, as well as fast growth rates and the capacity to accumulate or secrete metabolites. Microalgal biorefineries allow the transformation of biomass into the production of fuels, food, feed, chemicals, polymers and value-added ingredients [92].

Thus, the use of these microorganisms in carbon sequestration processes combines the treatment of polluting compounds with the production of consumables in a variety of forms. Table 4 shows some potential uses for the bioproducts obtained from microalgal biorefineries and formation by the biological conversion of CO_2 in photobioreactors.

Activity	Application	Reference
Nutraceutical, antimicrobial, anti-inflammatory	Nutritional supplement, antiproliferative, combat infections and diseases.	[93,94]
Antioxidant, natural pigment	Supplement and food ingredient for humans, feeding of fish and shellfish.	[93,95]
Biofuels	Natural gas production in fermenters by the digestion of biomass for obtain biodiesel.	[96,97]
Fertilizers	Use of the biomass as a source of nitrogen and phosphorous in tillable land.	[98]
High-value molecules	Chlorophyll-*a*, phycocyanin, β-carotene, γ-linolenic acid, eicosapentaenoic acid and stable biochemical isotopes.	[99]
Anticancer and antitumor	Antiproliferative. Inducing G1 inhibition in post-gastric carcinoma cells.	[94]
Chemical industry	Volatile organic compounds.	[100]

Table 4. Potential uses for bioproducts obtained from microalgal biorefineries

Microalgae possess a versatile metabolic capacity that can be transformed into valuable products through various processing routes. Some microalgae species as *Chlorella, Chlamydomonas, Dunaliella, Scenedesmus,* and *Tetraselmis* have a high carbohydrate content (37–55%) that mainly comes from starch in chloroplasts and cellulose cell walls [101]. Carbohydrate-rich microalgal biomass were evaluated for bioethanol production and were found to provide good yields [102].

The lipid profile of microalgae shows values of 2–77% depending on species and growth conditions. Microalgae lipids are classified into two groups, one for transformation in biofuel and one for food supplements, with carbon numbers of between 14–20 and 20 carbons respectively. Microalgae have a promising future, with production of eicosapentaenoic acid and docosahexaenoic acid as the main product. The species of microalgae producing omega-3 polyunsaturated fatty acids are mainly *Bacillariophyta, Chlorophyta, Cryptophyta, Haptophyta, Heterokontophyta* and *Rhodophyta* [103,104].

Proteins are among the main constituents of microalgae, at proportions of 50–70% depending on species, and they are an important product of microalgae biorefineries. Microalgal proteins can be used in human or animal nutrition (from aquaculture to farm animals). However, some microalgae contain toxic proteins, so analytical analyses need to be performed [105]. Nutritional and toxicological evaluations have demonstrated that microalgal biomass offers a valuable feed supplement or substitute for conventional animal feed sources [106].

Microalgae are known to be a good source of pigments and bioactive compounds. Chlorophylls, phycobilins and carotenoids are molecules with a high added value that can be obtained from *Porphyridium cruentum, Synechococcus* sp. and *Chlorella* and used in the chemical industry. Rodrigues et al. [107] showed that *Phormidium autumnale* has potential for the production of carotenoids. Sensitivity analysis showed the possibility of obtaining 107,902.5 kg/year of total carotenoids at the industrial scale. Symplostatin and curacin A have been isolated from the cyanobacteria *Symploca hydnoides* and *Lyngbya majuscula* respectively. These compounds exhibited cytotoxicity against a human carcinoma cell line [108].

The microalgae biorefineries industry promises much from the economic point of view. Global annual sales of beta-1,3-glucan from *Chlorella* sp. are in excess of USD$38 billion [105]. Moreover, phycobiliproteins present in cyanobacteria and some algae used to develop compounds for the pharmaceutical industry, represented a market of about USD$6–11 million with prices that varied from USD$3–25 mg^{-1} [105]. Considering that Kenekar and Deodhar [109] reported a phycocyanin yield of 0.071 gL^{-1} in *Geitlerinema sulphureum* culture, a photobioreactor with 100 L could generate a profit of approximately USD$177,500 [109]. Microalgae biomass produces more than 5,000 tons of dried mass/year with an annual revenue greater than USD$1.25 billion, not including processed products, demonstrating the potential of this type of biotechnological process [99]. Despite the promising conditions for the production of microalgae biomass and bioproducts, the industrial-scale development is currently a long way from the high profits available in theory. This is due to the lack of methods and photobioreactors that can produce large enough quantities to supply the market [12].

http://dx.doi.org/10.5772/59969

Finally, microalgae cells can produce methane. Sialve et al. [110] showed methane production values from anaerobic digestion in microalgae biomass in the range of 0.09–0.54 L CH_4/g volatile solids [110]. Furthermore, compounds such as non-methane hydrocarbon, organohalogens, and aldehydes are continuously being formed and released from the liquid phase of photobioreactors. The production of renewable polymers is an emerging industrial field [4].

Therefore, microalgal biotechnology can be seen as a promising scientific tool in the near future and microalgae biorefineries have the potential to solve some of the environmental, nutritional and pharmaceutical problems afflicting society.

9. Final considerations

Most research into microalgae biorefineries has been undertaken at the laboratory or pilot scale, and the number of full-scale studies is limited. Large-scale microalgae processes have been developed mainly using open photobioreactors. Some successful initiatives have been carried out in closed systems, but the closed systems need to operate at a large scale, to overcome the many drawbacks. At a large scale, algal growth conditions need to be closely controlled. The processes can be economical when using inexpensive sources of CO_2 from flue gas emissions, wastewaters, and/or with the extraction of bioproducts for industrial use.

Finally, many companies are investing in biotechnology, increasing spending on the production systems in order to obtain microalgal biofuel and high value-added bioproducts. Although at present there is no consensus on the criteria for the large scale development of photobioreactors for microalgae cultivation. Conventional configurations of closed systems, and hybrid photobioreactors are being employed and constantly improved for use at the industrial scale.

Author details

Eduardo Jacob-Lopes[1*], Luis Guillermo Ramírez Mérida[2], Maria Isabel Queiroz[3] and Leila Q. Zepka[1]

*Address all correspondence to: jacoblopes@pq.cnpq.br

1 Food Science and Technology Department, Federal University of Santa Maria, UFSM, Santa Maria, RS, Brazil

2 Department of Biology, University of Carabobo, UC, Avenue Universidad, Valencia, Edo. Carabobo, Venezuela

3 School of Chemistry and Food, Federal University of Rio Grande, FURG, Eng. Alfredo Huch, CEP Rio Grande, RS, Brazil

References

[1] Song C. Global challenges and strategies for control, conversion and utilization of CO_2 for sustainable development involving energy, catalysis, adsorption and chemical processing. Catalysis Today 2006; 115(1–4):2-32.

[2] United Nations. Kyoto Protocol to the United Nations Framework Convention on Climate Change. 1998. http://unfccc.int/resource/docs/convkp/kpeng.pdf. (accessed 28 March 2014).

[3] Ramírez-Mérida L., Zepka Q.L., Jacob-Lopes E. Microalgae and cyanobacteria: application in medicine. Revista Electrónica PortalesMedicos.com 2014; 9(4):149.

[4] Jacob-Lopes E., Teixeira Franco T. Microalgae-based Systems for Carbon Dioxide Sequestration and Industrial Biorefineries. In: Momba M. and Bux F. (ed.) Biomass. Croatia: Sciyo; 2010. p135-146.

[5] Vanthoor-Koopmans M., Wijffels R., Barbosa M., Eppink M. Biorefinery of microalgae for food and fuel. Bioresource Technology 2013; (135):142-149.

[6] Singh R.M., Sharma S. Development of suitable photobioreactor for algae production – A review. Renewable & Sustainable Energy Reviews. 2012;16(4):2347–2353.

[7] Ramírez-Mérida L., Zepka Q.L., Jacob-Lopes E. Photobioreactor: tool for mass cultivation of cyanobacteria. Ciencia y Tecnología 2013; 6(2):9-19.

[8] Thitakamol B., Veawab A., Aroonwilas A. Environmental impacts of absorption based CO_2 capture unit for post-combustion treatment of flue gas from coal-fired power plant. International Journal of Greenhouse Gas Control 2007; 1:318-342.

[9] Le Quéré C., Raupach M.R., Canadell J.G., Marland G., Bopp L., Ciais P., Conway T.J., Doney S.C., Feely R.A., Foster P., Friedlingstein P., Gurney K., Houghton R.A., House J.I., Huntingford C., Levy P.E., Lomas M.R., Majkut J., Metzl N., Ometto J.P., Peters G.P., Prentice I.C., Randerson J.T., Running S.W., Sarmiento J.L., Schuster U., Sitch S., Takahashi T., Viovy N., van der Werf G.R., Woodward F.I. Trends in the sources and sinks of carbon dioxide. Nature Geoscience 2009; 2:831–6.

[10] Kadam K.L. Power plant flue gas as a source of CO_2 for Microalgae cultivation: economic impact of different process options. Energy Conversion and Management 1997; 38(Suppl.):505–10.

[11] Ho S-H., Li P-J., Liu C-C., Chang J-S. Bioprocess development on microalgae-based CO_2 fixation and bioethanol production using *Scenedesmus obliquus* CNW-N. Bioresource Technology 2013, (145):142-149.

[12] Ramírez-Mérida, L.G., Zepka, L.Q., Jacob-Lopes, E. Why does the Photobioreactors Fail? Journal of Bioprocessing & Biotechniques 2015; 5:7.

[13] Jacob-Lopes, E., Revah, S., Hernández, S., Shirai, K., Franco, T.T., 2009. Development of operational strategies to remove carbon dioxide in photobioreactors. Chemical Engineering Journal 153, 120-126.

[14] Jacob-Lopes, E., Lacerda, L.M.C.F., Franco, T.T., 2008. Biomass production and carbon dioxide fixation by *Aphanothece microscopica Nägeli* in a bubble column photobioreactor. Biochem. Eng. J. 40, 27-34.

[15] Skjanes K, Lindblad P, Muller J. BiOCO_2 — a multidisciplinary, biological approach using solar energy to capture CO_2 while producing H_2 and high value products. Biomolecular Engineering 2007; 24:405–13.

[16] Wang B., Li Y.Q., Wu N., Lan C.Q. CO_2 bio-mitigation using microalgae. Applied Microbiology and Biotechnology 2008; 79:707–718.

[17] Van Den Hende S., Vervaeren H., Boon, N. Flue gas compounds and microalgae: (bio-) chemical interactions leading to biotechnological opportunities. Biotechnology Advances 2012; 6:1405–1424.

[18] Kumar K, Dasgupta C.N., Nayak B., Lindblad P., Das D. Development of suitable photobioreactors for CO_2 sequestration addressing global warming using green algae and cyanobacteria. Bioresource Technology 2011; 102:4945–4953.

[19] Kvamsdal H.M., Haugen G., Svendsen H.F. Flue-gas cooling in postcombustion capture plants. Chemical Engineering Research and Design 2011; 89:1544–1552.

[20] Chiu S.Y., Kao C.Y., Huang T.T., Lin C.J., Ong S.C., Chen C.D., Chang J.S., Lin C.S. Microalgal biomass production and on-site bioremediation of carbon dioxide, nitrogen oxide and sulfur dioxide from flue gas using *Chlorella* sp. cultures. Bioresource Technology 2011; 102:9135–9142.

[21] Xu Y., Isom L., Hanna M.A. Adding value to carbon dioxide from ethanol fermentations. Bioresource Technology 2010; 101:3311–3319.

[22] Algae Floating Systems technologies. AFS BioOil. http://www.afsbiooil.com/company-history.html (accessed 18 August 2014).

[23] Algae Floating Systems, Inc. AFS Biofarm™. http://www.algaefloatingsystems.com/company-3.html (accessed 18 August 2014).

[24] Seambiotic Ltd. http://www.seambiotic.com/home-2/ (accessed 18 August 2014).

[25] Aeon Biogroup. http://www.aeonbiogroup.com/ (accessed 18 August 2014).

[26] IGV Biotech. http://www.igv-gmbh.de/en/home/ (accessed 18 August 2014).

[27] Algenol Biofuels. http://www.algenol.com/ (accessed 18 August 2014).

[28] Becker E.W. Microalgae as a source of protein. Biotechnology Advances 2007; 25:207–210.

[29] Zepka Q.L., Jacob-Lopes E., Goldbeck R., Queiroz M.I. Production and biochemical profile of the microalgae *Aphanothece microscopica Nageli* submitted to different drying conditions Chemical Engineering and Processing 2008; 47:1305–1310.

[30] Zhu L.D., Hiltunen E., Antila E., Zhong J.J, Yuan Z.H., Wang Z.M. Microalgal biofuels: Flexible bioenergies for sustainable development. Renewable and Sustainable Energy Reviews 2014; 30:1035–1046.

[31] Bergman L., Ran D.G.A. Cyanobacterial–plant symbioses: signalling and development. In: Herrero A., Flores E. (ed.) The Cyanobacteria, Molecular Biology, Genomics and Evolution. Norfolk, UK: Caister Academic Press; 2008. p447–473.

[32] Nelson D.L., Cox M.M. Lehninger principles of biochemistry. (4 ed.) New Jersey: W.H. Freeman & Co.; 2005.

[33] Calvin M. 40 years of photosynthesis and related activities. Photosynthesis Research 1989; 21:3-16.

[34] Iverson T.M. Evolution and unique bioenergetic mechanisms in oxygenic photosynthesis. Current Opinion in Chemical Biology 2006; 10:91-100.

[35] Brennan L., Owende P. Biofuels from microalgae – a review of technologies for production, processing, and extractions of biofuels and co-products. Renewable & Sustainable Energy Reviews 2010; 14:557–577.

[36] Ravelonandro P.H., Ratianarivo D.H., Joannis-Cassan C., Isambert A., Raherimandimby M. Influence of light quality and intensity in the cultivation of *Spirulina platensis* from Toliara (Madagascar) in a closed system. Journal of Chemical Technology and Biotechnology 2008; 83:842–848.

[37] Jaiswal P., Kashyap A. Isolation and characterization of mutants of two diazotrophic cyanobacteria tolerant to high concentrations of inorganic carbon. Microbiological Research 2002; 157:83-91.

[38] Marcus Y. Distribution of inorganic carbon among its component species in cyanobacteria: Do cyanobacteria in fact actively accumulate inorganic carbon? Journal of Theoretical Biology 1997; 185:31-45.

[39] Badger M.R., Price G.D. CO_2 concentration mechanisms in cyanobacteria: Molecular components, their diversity and evolution. Journal of Experimental Botany 2003; 54(383):609-622.

[40] Cuaresma M., Garbayo I., Vega J.M., Vílchez C. Growth and photosynthetic utilization of inorganic carbon of the microalga *Chlamydomonas acidophila* isolated from Tinto river. Enzyme and Microbial Technology 2006; 40:158-162.

[41] Kumar A., Ergas S., Yuan X., Sahu A., Zhang Q., Dewulf J., Malcata F.X., van Langenhove H. Enhanced CO_2 fixation and biofuel production via microalgae: recent developments and future directions. Trends in Biotechnology 2010; 28:371–80.

[42] Welsh D.T., Bartoli M., Nizzoli D., Castaldelli G., Riou S.A., Viaroli P. Denitrification, nitrogen fixation, community primary productivity and inorganic-N and oxygen fluxes in an intertidal *Zostera noltii* meadow. Marine Ecology Progress Series 2000; 208:65–77.

[43] Hsieh C-H., Wu W-T. Cultivation of microalgae for oil production with a cultivation strategy of urea limitation. Bioresource Technology 2009; 100(17):3921–3926.

[44] Çelekli A., Yavuzatmaca M., Bozkurt H. Modeling of biomass production by *Spirulina platensis* as function of phosphate concentrations and pH regimes. Bioresource Technology 2009; 100(14):3625–3629.

[45] Ben-Amotz A., Avron A. The biotechnology of mass culturing *Dunaliella* for products of commercial interest. In: Cresswell R.C., Rees T.A.V., Shah N. (ed.) Algal cyanobacterial biotechnology. Essex: Longman Scientific & Technical; 1989 p91–114.

[46] Stephenson A.L., Dennis J.S., Howe C.J., Scott S.A., Smith A.G. Influence of nitrogen-limitation regime on the production by *Chlorella vulgaris* of lipids for biodiesel feedstocks. Biofuels 2010; 1:47–58.

[47] Blanken W., Cuaresma M., Wijffels R., Janssen M. Cultivation of microalgae on artificial light comes at a cost. Algal Research 2013; 2:333–340

[48] Yeh N.C., Chung J.P. High-brightness LEDs-energy efficient lighting sources and their potential in indoor plant cultivation. Renewable & Sustainable Energy Reviews 2009; 13:2175–2180

[49] Wang C.Y., Fu C.C., Liu Y.C. Effects of using light-emitting diodes on the cultivation of *Spirulina platensis*. Biochemical Engineering Journal 2007; 37:21–25.

[50] Melnicki M., Pinchuk G., Hill E., Kucek L., Stolyar S., Fredrickson J., Konopka A., Beliaev A. Feedback-controlled LED photobioreactor for photophysiological studies of cyanobacteria. Bioresource Technology 2013; 134:127–133.

[51] Katsuda T., Shimahara K., Shiraishi H., Yamagami K., Ranjbar R., Katoh S. Effect of flashing light from blue light emitting diodes on cell growth and astaxanthin production of *Haematococcus pluvialis*. Journal of Bioscience and Bioengineering 2006; 102:442–446.

[52] Zhao B., Su Y. Process effect of microalgal-carbon dioxide fixation and biomass production: A review. Renewable and Sustainable Energy Reviews 2014; 31:121–132.

[53] Janssen M., Tramper J., Mur L.R., Wijffels R.H. Enclosed out door photobioreactors: light regime, photosynthetic efficiency, scale-up, and future prospects. Bioengineering and Biotechnology 2003; 81:193–210.

[54] Jacob-Lopes E., Gimenes Scoparo C.H., Ferreira Lacerda L.C.M., Teixeira Franco T. Effect of light cycles (night/day) on CO_2 fixation and biomass production by microalgae in photobioreactors. Chemical Engineering and Processing 2009; 48:306–310.

[55] Carvalho A.P., Meireles L.A., Malcata F.X. Microalgal reactors: a review of enclosed system designs and performances. Biotechnology Progress 2006; 22:1490–1506.

[56] Schulze P.S.C., Barreira L.A., Pereira H.G.C., Perales J.A., Varela J.C.S. Light emitting diodes (LEDs) applied to microalgal production. Trends in Biotechnology 2014; 32(8): 422-430.

[57] Xue S., Zhang Q., Wu X, Yan C, Cong W. A novel photobioreactor structure using optical fibers as inner light source to fulfill flashing light effects of microalgae. Bioresource Technology 2013; 138:141–147

[58] Jacob-Lopes E., Zepka L.Q., Queiroz M.I. Cyanobacteria and carbon sequestration. In: Sharma N., Rai A and Stal L. (ed.) Cyanobacteria an economic perspective. Chichester:Wiley Blackwell; 2014. p65-71.

[59] Singh R.M., Sharma S. Development of suitable photobioreactor for algae production – A review. Renewable & Sustainable Energy Reviews 2012; 16(4):2347–2353.

[60] Dasgupta C.N., Gilbert J.J., Lindblad P., Heidorn T., Borgvang S.A., Skjanes K., Das D. Recent trends on the development of photobiological processes and photobioreactors for the improvement of hydrogen production. International Journal of Hydrogen Energy 2010; 35(19):10218–10238.

[61] Wang B., Li Y., Wu N., Lan C. CO_2 bio-mitigation using microalgae. Applied Microbiology and Biotechnology 2008; 79(5):707–18.

[62] Gudin C., Therpenier C. Bioconversion of solar energy into organic chemicals by microalgae. Advances in Biotechnological Processes 1986; 6:73–110.

[63] Molina Grima E., Belarbi E.H., Acién Fernández F.G., Robles Medina A., Chisti Y. Recovery of microalgal biomass and metabolites: process options and economics. Biotechnology Advances 2003; 20(7–8):491–515.

[64] Petrusevski B., Bolier G., Van Breemen A.N., Alaerts G.J. Tangential flow filtration: a method to concentrate freshwater algae. Water Research 1995; 29:1419–1424.

[65] Mohn F.H. Experiences and strategies in the recovery of biomass in mass culture of microalgae. In: Shelef G., Soeder C.J. (ed.) Algal biomass. Amsterdam: Elsevier; 1980. p547–71.

[66] MacKay D., Salusbury T. Choosing between centrifugation and cross flow microfiltration. Chemical Engineering Journal 1988; 477:45–50.

[67] Knuckey R.M., Brown M.R., Robert R., Frampton D.M.F. Production of microalgal concentrates by flocculation and their assessment as aquaculture feeds. Aquacultural Engineering 2006; 35:300–313.

[68] Heasman M., Diemar J., O' Connor W., Sushames T., Foulkes L., Nell J.A. Development of extended shelf-life microalgae concentrate diets harvested by centrifugation for bivalve molluscs—a summary. Aquaculture Research 2000; 31(8 – 9):637– 59.

[69] Dassey A., Theegala C. Harvesting economics and strategies using centrifugation for cost effective separation of microalgae cells for biodiesel applications Bioresource Technology 2013; 128:241-245.

[70] Coward T., Lee J., Caldwell G. Harvesting microalgae by CTAB-aided foam flotation increases lipid recovery and improves fatty acid methyl ester characteristics Biomass and Bioenergy 2014; 67:354-362.

[71] Muñoz R., Guieysse B. Algal-bacterial processes for the treatment of hazardous contaminants: a review. Water Research 2006; 40(15):2799–815.

[72] Bruton T., Lyons H., Lerat Y., Stanley M., Rasmussen M.B. A review of the potential of marine algae as a source of biofuel in Ireland. Dublin: Sustainable Energy Ireland; 2009. p. 88.

[73] Uduman N., Qi Y., Danquah M.K. Forde G.M., Hoadley, A. Dewatering of microalgal cultures: a major bottleneck to algae-based fuels. Journal of Renewable and Sustainable Energy 2010; 2:012701–012715.

[74] Garg S., Wang L., Schenk P. Effective harvesting of low surface-hydrophobicity microalgae by froth flotation. Bioresource Technology 2014; 159:437–441.

[75] Nurra C., Torras C., Clavero E, Ríos S., Rey M, Lorente E., Farriol X., Salvadó J. Biorefinery concept in a microalgae pilot plant. Culturing, dynamic filtration and steam explosion fractionation. Bioresource Technology 2014; 163:136-142

[76] Jacob-Lopes E, Teixeira Franco T. From oil refinery to microalgal biorefinery. Journal of CO_2 Utilization 2013; 2:1-7.

[77] Guldhe A., Singh B., Rawat I., Ramluckan K., Bux F. Efficacy of drying and cell disruption techniques on lipid recovery from microalgae for biodiesel production. Fuel 2014; 128(15):46-52.

[78] Goettel M., Eing C., Gusbeth C., Straessner R., Frey W. Pulsed electric field assisted extraction of intracellular valuables from microalgae. Algal Research 2013; 2(4): 401-408.

[79] Frey W., Gusbeth C., Schwartz T. Inactivation of *Pseudomonas putida* by pulsed electric field treatment: a study on the correlation of treatment parameters and inactivation efficiency in the short-pulse range. Journal of Membrane Biology 2013; 246(10): 769-781.

[80] Ranjan A., Patil C., Moholkar V.S. Mechanistic assessment of microalgal lipid extraction. Industrial & Engineering Chemistry Research 2010; 49(6):2979–2985.

[81] Ehimen E., Sun Z., Carrington G. Use of Ultrasound and Co-Solvents to Improve the In-Situ Transesterification of Microalgae Biomass. Procedia Environmental Sciences 2012; 15:47-55.

[82] Fu C-C., Hung T-C., Chen J-Y., Su C-H., Wu W-T. Hydrolysis of microalgae cell walls for production of reducing sugar and lipid extraction. Bioresource Technology 2010; 8750

[83] Sathish A., Sims R. Biodiesel from mixed culture algae via a wet lipid extraction procedure. Bioresource Technology 2012; 118:643–647.

[84] Richmond A. Handbook of microalgal culture: biotechnology and applied phycology. Blackwell Science Ltd; 2004.

[85] Belarbi H., Molina E., Chisti Y. A process for high and scaleable recovery of high purity eicosapentaenoic acid esters from microalgae and fish oil. Enzyme and Microbial Technology 2000; 26:516–29.

[86] Araya B., Gouveia L., Nobre B., Reis A., Chamy R., Poirrier P. Evaluation of the simultaneous production of lutein and lipids using a vertical alveolar panel bioreactor for three *Chlorella* species. Algal Research, In Press, Corrected Proof, Available online 4 July 2014

[87] Du Y., Schuur B., Samorì C., Tagliavini E., Brilman DWF. Secondary amines as switchable solvents for lipid extraction from non-broken microalgae. Bioresource Technology 2013; 149: 253-260

[88] Amin Hejazi M., Wijffels R.H. Milking of microalgae. Trends in Biotechnology 2004; 22(4):189–94.

[89] Abrahamsson V., Rodriguez-Meizoso I., Turner C. Determination of carotenoids in microalgae using supercritical fluid extraction and chromatography. Journal of Chromatography A 2012; 1250:63-68.

[90] Solana M, Rizza C.S., Bertucco A. Exploiting microalgae as a source of essential fatty acids by supercritical fluid extraction of lipids: Comparison between *Scenedesmus obliquus, Chlorella protothecoides* and *Nannochloropsis salina.* The Journal of Supercritical Fluids 2014; 92:311-318.

[91] Hammann S., Tillmann U., Schröder M., Vetter W. Profiling the fatty acids from a strain of the microalgae *Alexandrium tamarense* by means of high-speed counter-current chromatography and gas chromatography coupled with mass spectrometry. Journal of Chromatography A 2013; 1312:93-103.

[92] Christenson L., Sims R. Production and harvesting of microalgae for wastewater treatment, biofuels, and bioproducts. Biotechnology Advances 2011; 29:686–702.

[93] Pulz O., Gross W. Valuable products from biotechnology of microalgae. Applied Microbiology and Biotechnology 2004; (65):635–48.

[94] Sheih I.C., Fang T., Wu TK., Lin PH. Anticancer and antioxidant activities of the peptide fraction from algae protein waste. Journal of Agricultural and Food Chemistry 2010; 58 (2):1202–1207.

[95] Becker W. Microalgae in human and animal nutrition. In: Richmond A, (ed.) Handbook of microalgal culture. Oxford: Blackwell Publishing; 2004. p312–51.

[96] Yen H.W., Brune D.E. Anaerobic co-digestion of algal sludge and waste paper to produce methane. Bioresource Technology 2007; 98:130-134.

[97] Miao X., Wu Q. Biodiesel production from heterotrophic microalgal oil. Bioresource Technology 2006; 97:841-846.

[98] Chae S.R., Hwang E.J., Shin H.S. Single cell protein production of *Euglena gracilis* and carbon dioxide fixation in an innovative photo-bioreactor. Bioresource Technology 2006; 97:322-329.

[99] Spolaore P., Cassan C.J., Duran E., Isambert A. Commercial applications of microalgae. Journal of Bioscience and Bioengineering 2006; 101:87-96.

[100] Muñoz, J., Mudge, S.M., Sandoval, A. Effects of ionic strength on the production of short chain volatile hydrocarbons by *Dunaliella salina* (Teodoresco) 2004; Chemosphere 54: 1267-1271.

[101] Dragone, G., Fernandes, B.D., Abreu, A.P., Vicente, A.A., Teixeira, J.A., 2011. Nutrient limitation as a strategy for increasing starch accumulation in microalgae. Applied Energy 2011; 88(10):3331–3335.

[102] Ho S-H., Huang S-W., Chen C-Y., Hasunuma T., Kondo A., Chang J-S. Bioethanol production using carbohydrate-rich microalgae biomass as feedstock. Bioresource Technology 2013; 135:191–198

[103] Ryckebosch E., Bruneel C., Muylaert K., Foubert I. Microalgae as an alternative source of omega-3 long chain polyunsaturated fatty acids. Lipid Technology 2012; 24(6):128–130.

[104] Ryckebosch E., Bruneel C., Termote-Verhalle R., Goiris K., Muylaert K., Foubert I. Nutritional evaluation of microalgae oils rich in omega-3 long chain polyunsaturated fatty acids as an alternative for fish oil. Food Chemistry 2014; 160:393–400.

[105] Yaakob Z., Ali E., Zainal A., Mohamad M., Takriff M.S. An overview: biomolecules from microalgae for animal feed and aquaculture. Journal of Biological Research-Thessaloniki 2014; 21:6.

[106] Harun R., Singh M., Forde G.M., Danquah M.K. Bioprocess engineering of microalgae to produce a variety of consumer products. Renewable & Sustainable Energy Reviews 2010; 14:1037–1047.

[107] Rodrigues D., Flores E., Barin J, Mercadante A, Jacob-Lopes E, Zepka Q.L. Production of carotenoids from microalgae cultivated using agroindustrial wastes. Food Research International, In Press, Corrected Proof, Available online 4 July 2014.

[108] Gerwick W.H., Tan L., Sitachitta N. Nitrogen-containing metabolites from marine cyanobacteria. In Cordell, G.A. (ed.) The Alkaloids. San Diego: Academic Press; 2001. p75-184

[109] Kenekar A., Deodhar M. Effect of Varying Physicochemical Parameters on the Productivity and Phycobiliprotein Content of Indigenous Isolate *Geitlerinema sulphureum*. Biotechnology 2013; 12(3):146-154.

[110] Sialve B., Bernet N., Bernard O. Anaerobic digestion of microalgae as a necessary step to make microalgal biodiesel sustainable. Biotechnology Advances 2009; 27:409–416.